T0100407

WEARABLE TECHNOLOGY

www.royalcollins.com

KEVIN CHEN

WEARABLE TECHNOLOGY

Books Beyond Boundaries

ROYAL COLLINS

Wearable Technology

Kevin Chen

First published in 2021 by Royal Collins Publishing Group Inc.
Groupe Publication Royal Collins Inc.
BKM Royalcollins Publishers Private Limited

Headquarters: 550-555 boul. René-Lévesque O Montréal (Québec) H2Z1B1 Canada
India office: 805 Hemkunt House, 8th Floor, Rajendra Place, New Delhi 110 008

ISBN: 978-1-4878-0487-9

To find out more about our publications, please visit www.royalcollins.com

PREFACE

In 2012 Google surprised the world by announcing the arrival of Google Glass, a radical advancement in wearable technology. The wearable phenomenon has since grown rapidly, spreading to different technologies and applications with increasing market penetration. Underpinning this growth has been an industrial revolution, to a certain extent, with a growing industry based on hardware devices and software platforms, structured on novel business models and related ecosystems.

Wearable fitness and medical devices have perhaps seen the largest uptake to date. From Fitbit to Apple Health, these apps and wearables undoubtedly have great potential to improve the health and well-being of those wearing these devices. Many such wearables are based on traditional product-based business models. Some devices, including Fitbit, use novel business models, going beyond the traditional product-based value chain to new chains based on the data provided by users and the technology. Makers of medical wearables have begun partnering with insurance companies to gain additional value from the data they collect.

The potential of wearable technology is significant. From advertising to tourism, medicine to gaming, there is hardly a sector that would not be impacted by this technology. Current applications of this technology include smartwatches, activity trackers, virtual reality devices and medical applications. These are only the first fruits of a revolution that may lead to dramatic shifts in the way businesses interact with customers, the way in which a state relates to its citizens, and the way people interact with each other.

To date there have been fewer applications in the fields of advertising, tourism, and e-commerce. This can be seen as the result of barriers in those business ecosystems—including the lack of standardized data collection, analysis, or customer feedback in these sectors. However, a significant portion of this book will examine potential future business models in these sectors, drawing on current market trends and scanning the horizon for future technology that may influence other business areas.

According to IDC, the market growth of wearable technology is faster than any other segment of consumer electronics globally. The worldwide sales of wearables in 2014 was three times that of 2013, with over 192 million units sold. It is estimated that this figure may reach 1.26 billion units by 2019. Global sales may reach as high as USD 27.9 billion by 2019. In IDC's report on wearables for the first quarter of 2015, the total shipment volume was 11.4 million units, reflecting a 200% year-on-year increase and representing the eighth consecutive quarter of continuous growth.

According to a wearable technology market forecast by NPD Display Search, the Chinese mainland represents the largest potential market for wearables. According to the same study, the enormous appetite of the Chinese market has made a significant global impact on the design, specifications, cost, and pricing for many consumer electronics products—and this applies equally to the wearable technology industry. With its vast market demands, China may take the lead in developing business models with unique Chinese characteristics.

These forecasts merely represent the tip of the iceberg in terms of the potential of the entire wearable technology industry. They are largely based on currently existing products such as smartwatches and activity-tracking wristbands. But the future of the wearable market includes both devices worn on the human body and those that can be implanted inside.

Wearable devices are largely thought of as those worn externally on the body. With the emergence of wearable technology, familiar branded wearables such as the Apple Watch and Google Glass were the leading products of this type, and products from both companies continue to lead along with many other smartwatches and activity trackers offered by a variety of businesses—from large corporations to tech start-ups. But watches and trackers are only scratching the surface of the world of wearable technology. In the future it is probable we will see embedded wearables, to the point where wearing a device is no longer necessary. Indeed, in terms of the entire industry, the market capacity of smartwatches and activity trackers is rather small. The nascent markets for, clothing, shoes, or accessories all have much greater potential. The actual market capacity of wearables is well beyond our current understanding.

In this book I will look into the trajectory of wearable technology against the backdrop of global developments as well as many popular applications. This will build on real-life case studies and future projections as I analyze and explore the business models of wearable technology. I hope this book will help businesses, start-ups, and individuals interested in this industry find a good entry point, and in doing so receive better returns on investments. Given the fast pace of development in this sector, this cannot be an exhaustive review of potential products or business models, but is focused on areas that have been selected to represent the entire wearable technology industry.

PROLOGUE

"We are right at the beginning of a technology revolution... People assume the technology tomorrow will be similar to that today. What they do not know is the fact that technology is exploding right in front of our eyes, and this will change everything."

—Lee Silver
Professor of Genetics at Princeton University

Over the past seventy years, the world has experienced an unprecedented revolution in information technology, shifting from the world economy of the industrial era to an internet-based new global economy. The number of inventions and technological innovations in the past 20 years exceeds the entirety of the previous two to three centuries. This wave of technology has swept over our lives, transforming livelihoods and lifestyles.

At the time of writing, the rise of the smart phone era is driving what may be seen as the wearable device age, with information collection and presentation based on a smart end-device closely associated with the human body. These devices will be naturally incorporated into our body, and blend into and rebuild our lives. Future economic development is likely to show a very different pattern. When the people at the center of the economy are entirely "hijacked," it is inevitable that the world economy will need to go through a radical transformation in which intelligent technology centered around wearables will lead the way.

Following the introduction of the internet, we have seen significant changes in the way we live our lives and conduct our business. Looking ahead, wearable technology will build on this – causing a fundamental change in livelihoods, lifestyles, business-customer interactions, public management of services, and the working of the economy itself.

1.1   A future economy led by Smart Technology

In the past 20 to 30 years, in the study of information networks and in bioscience there have been some advances that may seem, at first glance, to be rather insignificant compared to advances like the discovery of the Higgs boson. However, what seem like pebbles dropped into the ocean of knowledge are causing ripples, which could bring about major waves in the near future. Without knowing much of the bigger picture, we have captured many of the opportunities presented at every point of the technological development.

We have been amazed by advancements in bio-science over the years. Consider how our understanding of fertility has advanced. On July 25, 1978, Louise Brown was born in England, the first human baby born after conception by in vitro fertilization (IVF). The news of the birth sent shock waves throughout the world. Many people were worried, with words like "Frankenstein" used to describe this advancement. And yet Louise Brown grew up healthily and is happily married with her own children now. Two decades later, in 1997, a sheep named Dolly was successfully cloned by Ian Wilmut, a British embryologist. The birth of Dolly caused controversy, as it suggested the possibility of advanced mammals, including humans, being cloned. Neither the invention of IVF nor the cloning of Dolly the sheep have led yet to radical change in the way the world lives – but when these advances are coupled with smart technology, new doorways may open to treatments or changes in the way food systems work.

Smart phones have fundamentally changed our social habits, transforming our lives in many ways. The way we shop, what we read, and how we communicate have all been affected. Our working lives have been changed. The new technologies offer ways to make things easier or work more efficiently. Mobile phones are no longer tools for communication alone. Instead, they have changed our life, our entertainment consoles, and our supermarkets.

The emergence of wearable technology will continue to change modern life as we know it, gradually replacing the smart phone to further overturn the old ways of life. Humans will enter what may be seen as a new epoch—the true age of "Big Intelligence."

Unlike smart phones, wearable devices will free our hands totally. Human-machine interaction will shift gradually to voice interaction, or even brain wave interaction through subconscious thoughts. If you consider yourself "tied down" today by a smart phone that is the center of your life, the emergence of wearables offers the potential to change this. Wearables could release us from the black hole information paradox and make digital information more human-centered and human-serving through advanced communication technology. The wearables of tomorrow will not only act as your personal assistant but could also be your private nurse, monitoring your heart and blood pressure, and contacting your doctor when there is an emergency.

Through DNA testing, today's medical profession is moving toward Precise Medicine. The aim is to fundamentally decode human diseases and predict potential health problems. With the help of wearables, both doctors and patients would have access to better data about their movement and other measures of health, allowing for the development of better, more effective medical treatments.

Bio-technology offers the potential to treat many diseases that are now incurable, including some cancers, Alzheimer's disease and more. Advances in 3D printing technology could enable the printing of human organs on demand to replace failing ones, enabling increased longevity and removing the need for organ donors. Likewise, human cells could be printed in order to replace and eliminate cancer cells.

1.1.1 The Eight Key Smart Future Trends

From smart phones to smart meters, smart transport systems to smart homes, it seems like wherever you may look the future is "smart." Knowing the road ahead in this changing landscape could enable individuals, businesses, and governments to better prepare for the future.

The first personal computer was launched in the 1970s. Since then, increasing computer power together with radical advances like the internet and genetic engineering have led to a second technological revolution, one driven by information technology. We can no longer doubt that human society is approaching a future economy and social system dominated by smart technology. This can be called a Smart Future. Here we will view and explore eight trends that underpin this smart revolution.

1. *Mobile Economy*

Mobile phone use is growing across the globe at an annual growth rate of 4%. By the year 2020, the population subscribed to mobile services was estimated to make up nearly 60% of the global total, a significant increase from 50% in 2014.[1]

There has been a gradual shift from traditional PC-based internet use to internet access via mobile devices. This includes both smartphones and wearables. In many ways, a brand new economic model has emerged, moving the entire market from one based on landline infrastructure to one based on mobile devices owned by consumers.

With various apps available for download to these devices, more users are receiving personalized services from online business platforms daily. It is forecast that in the coming 20 years, most business will be conducted on mobile devices, a new norm for the age of the Internet of Things.

2. *Sharing Economy*

The rise of the so-called sharing economy is perhaps shown best in the case of the ride sharing service Uber. Uber requires no fixed office, or contracts with job descriptions. It is flexible and profitable. Uber has changed people's lives as well as the way people work. The maximization of return is achieved through rational reallocation of idle resources. The most attractive attribute of the sharing economy model is its unbeatable flexibility. Anyone can do this and generate returns from it.

An Uber driver can be a university professor or a white-collar worker spending most of the day working behind a desk. Regardless of your routine job, as long as you have spare time and a vehicle that meets its requirements, you can be an Uber driver ready to earn some extra money. Research shows that the average education level of Uber drivers is rather high. Nearly half have bachelor's degrees and above, a much higher proportion than taxi drivers (18%) and above the average of the entire labor force (41%).

Instacart is a similar example. It gives the opportunity to earn money while shopping. As long as you have a smartphone, are over 18 years old and able to lift heavier items, , you can be part of Instacart, working at your own convenience. When you are shopping in a supermarket and you see a neighbor has placed an order on Instacart, you can take the order, shop for the person, and deliver it to their door. The pay is about USD 25 per hour.

Other players include Zipcar (car sharing), Airbnb (house sharing), BookCrossing (book sharing) and others, all providing platforms that enable consumers to rapidly identify the commodities they need at a much cheaper price. This model is the result of the unique data flow generated by the mobile technology that happens anywhere at any time.

3. *Boundless Computing*

In the coming future is an age of "complete intelligence" with cloud computing as the foundation. Cloud computing centers will gather and process information, while transmitting, storing and computing colossal volumes of complex data. Various screens will allow for communication and interface: ever-lighter smartphones and tablets, TV background walls, or even virtual holographic screens. Screens will be everywhere.

Boundless computing will change the way we work as well as our lifestyle. Traditionally, the average workday has been from nine to five, but as times change we will be presented with numerous job opportunities outside these traditional workplaces and hours, powered by the mobile network of information. In the US, only 35% of employees need to be physically present at the office during work hours. The rest of the workforce is already able to work at home or other venues, engaging in innovative businesses or providing customer services. Mobile smart devices enable instant communication with employers. The mobile network brings businesses and working environments together such as never before.

Division of labor also enters a new era with the Internet of Things. The essence of the division of labor is no longer about physical production processes, but rather in global information processing with wearables at the center. In the future Hong Kong may be the global data center for ophthalmology and US the center of pediatric data processing. Boundless computing will overcome geographic barriers and boost data processing platforms established around the world with countries leveraging their unique advantages in industrial data processing to serve the needs of the entire world.

4. Artificial Intelligence

In 2011 the supercomputer Watson competed on the US game show Jeopardy! against former champions, ultimately winning first place. In Japan, a female-looking humanoid robot dressed in kimono known as Aiko Chihira appeared at the information desk in the Mitsukoshi department store in Tokyo to provide customer services and introduce store events. This humanoid robot is developed by Toshiba to perform as a receptionist.

The advancement of artificial intelligence will have a profound impact on work and daily life. Particularly in the field of industrial manufacturing, intelligent robots or the use of mechanical arms may increase productivity considerably. Automated self-service is now readily available in many gas stations and supermarkets around the world. This overwhelming trend of process automation is gradually replacing human labor with robotics in all aspects of business life. This will undoubtedly pose a threat to human employment prospects.

This, however is merely the beginning. When AI integrates with Big Data, it will not only intelligently assist our lives, but will surround us. This seemingly unstoppable trend in the development of self-learning AI robotics will indeed impact and potentially pose a significant threat to human society.

5. Smart Living

Let us visualize a future lifestyle like this.

When you get in your car after a long day of work, your wearable device analyzes your energy level, mood, and the day's workload, understands your preferences at this moment, and automatically plays the perfect music for what you need, just as you start the engine. At the same time, it informs smart appliances at home to prepare the house for the right temperature and lighting. Your bath water will be ready if you want it and your healthy dinner is already being prepared.

You don't need to switch anything on when you arrive at home. Your smart home appliances know your habits and will prepare these options without your active input, using the data feed from your wearables. The choices only appear when you want them.

No human butler can provide a better level of service. Servants may get to know your food and clothing preferences over the years, but knowing what goes on in your mind is virtually impossible. And yet, these intelligent devices that can read you will create a perfect lifestyle for you. This scenario is on its way to becoming reality.

6. Regenerative Medicine

Regenerative medicine is a cutting-edge medical technique to repair human organs or tissue using stem cells. By growing tissues and organs in a laboratory and implanting them, it stimulates the body's own repairing mechanism to functionally heal damaged or irreparable organs or tissues in order to restore or establish normal function.

In 2006, Professor Shinya Yamanaka from Kyoto University and his team generated induced pluripotent stem cells (iPS cells). He was awarded the 2012 Nobel Prize in Physiology and Medicine. iPS cells could be used to grow all active human cells and tissues for teeth, nerves, retinas, cardiomyocytes, blood cells and liver cells, which could be implanted to restore damaged organs or tissues. Doctors could use this technology to grow stem cells from the patient's own cells, in order to discover the way a particular disease is developing (the 'pathogenesis'). Targeted treatment could then be developed to address that pathogenesis at the cellular level.

Regenerative medicine will become easier and more effective in the future. With the progress of bioprinting, 3D printing technology could be used to directly "print" human organs in the future. And 3D-printed nano-robots would serve as a safeguard inside human bodies, patrolling around the clock to identify cancer cells. They could even use a self-triggered shape-shifting function to treat or kill cancerous cells.

In a word, humans would no longer need to be afraid of losing the battle against a range of diseases. Medicine will likely be a leading sector in embracing the advances offered by smart technology, particularly with the arrival of the age of wearables, and will be radically changed as a result.

7. Thought Sharing

In the Bollywood movie *P.K.*, the alien P.K. doesn't communicate through spoken language or facial expression, but rather using handshakes and telepathy. This way of communication makes lying very difficult, but learning from one another quite simple.

Stephen Hawking, one of the greatest minds in recent history, developed the symptoms of amyotrophic lateral sclerosis (ALS) when he was 21. As the disease progressed, he became wheelchair bound, only able to move his two eyes and three fingers. By the end of his life, he employed the use of a novel device that used facial expressions to generate speech, adding to his expressiveness.

Many people might hold great enthusiasm about sharing thoughts or becoming telepathic. We can all recall moments during exams when we wish we had a photographic memory. When Ericsson conducted a global survey among smartphone users, 40% of them expressed the wish "to be able to communicate with others using thoughts via wearables." Two thirds of the participants even believed that by 2020 such communication would be quite common.

In terms of specific device requirements, 82% of consumers believed smartwatches that transmit tactile sensations would be mainstream by 2020. In second place, 72% of consumers wanted wearables that can control home devices like lamps and thermostats. Wearables with mind-reading functions for communication made up 69% of the support.

8. Forecast Monitoring
Data generated by search engines, GPS receivers, social media and other systems will increasingly impact the entire business world through the advancement of data analysis technology. Google, Facebook, and Twitter activity are used together with smartphones to provide extensive consumer behavior data. Big Data analytics build consumer profiles to precisely predict the shopping behavior of individuals. Supported by GPS, GIS, and SLAM (Simultaneous Localization And Mapping) systems in iOS or Android platforms, computers can also predict user movement by tracking the use of smartphones. Visual sensor systems work to monitor our daily activities around the clock, and privacy is pushed to the minimum. For that reason, futurist Patrick Tucker believes we will be living in a 'naked future' where everything is monitored and privacy will no longer exist.

Wearables enable even more accurate data collection. Commercial behavior is based on conscious and subconscious factors, with some purchasing decisions being made in an instant. Whoever masters the appropriate data first will be able to provide personalized services to customers first, and thereby win their attention and trust.

1.1.2   Consumer Models in the Future

1. Mainstreaming the personalized consumer model
Consumer behavior and preferences have shifted constantly throughout history. Today, in the 21st century, the mainstream consumer preference is a

bespoke and personalized one.

The launch of the iPhone 4 instantly reduced the distance between businesses and users, the first of a generation of increasingly advanced smartphones. All the commodities in the world are now at the consumer's fingertips and are centered on the individual. Mobile network technology has accelerated the flow of information and coverage in ways never seen before. Data exchange is freed from any geographic limits, allowing the maximum variation in user experiences. The Chinese firm Xiaomi leverages the social media platform Weibo to interact with users, making a user-oriented focus the driver of its business strategy, allowing customers to participate and voice their opinions, and eventually buy their own ideas. This user-oriented business model will be a part of the mainstream for a long time to come. It will generate a business pattern that is segmented, personalized, engaged, experiential and fast. Consumers will have a vast range of products available to purchase instantly. Everything will be only a click away. Consumers will start to redefine the value of things in their own ways.

2. More focus on product design

People are no longer satisfied with traditional values that focus solely on the functionality of products. The design and image of products has become a strong factor driving value in meeting modern consumer needs. Emotional value, a sense of attachment, is something many people have begun to look for in products as well.

Product design is presented in the appearance of a product, but it is also through design that the inner concept or the spirit of a product can be manifested. Consumers generally recognize the value of a product by not only selecting something for how it meets their needs, but for its design as well. In other words, basic material consumption is replaced by emotional consumption for many. The purchase for design is most noted among luxury commodities.

Personalized shopping is often reflected in purchasing by way of a bespoke design service. Products with a good design would be of lesser value if they are mass produced. Value is always increased with scarcity. Customers demand individualism, and personalized characteristics. If your product is special and uniquely made for me, the pressure of functionality on price and demand can be reduced.

3. Moving from a singular shopping pattern to a comprehensive one
People once moved from place to place in pursuit of one specific commodity that was needed. Now it is more common to shop in one place for all that is needed. Big shopping malls have become part of modern life. Online, at a click of a button, we not only have the world's commodities in front of us, but we can also have them delivered to our doorstep thanks to dramatically improved distribution networks. However, this is not yet at the level that can be called "comprehensive shopping." True comprehensive shopping means that a consumer completes a complex shopping process within a very short period of time.

A comprehensive shopping experience allows the integration of the purchasing of a range of commodities—from virtual products to physical, online to offline—cutting across needs at home or at work, all based on the use of mobile internet devices to enable the purchase of an extensive selection of products anywhere and at any time. One early example of a comprehensive shopping experience is provided by Alipay. Alipay has utilized its own platform to enable users to register for medical consultations, monitor queues, conduct payments, review results and more. Alipay will establish a comprehensive online platform to allow virtual mobile prescriptions, medicine deliveries, hospital transfers, and medical insurance reimbursements, as well as commercial insurance and damage claims. It will then further utilize its Big Data platform and cloud computing capability to work with wearable technology manufacturers, medical care institutions, and even government agencies to construct a healthcare management platform. It will facilitate a shift in healthcare from treatment to prevention in the future.

The future may be one with one-stop shops that deal with all our needs. This may even be the case for medical care, eliminating queues and reducing waiting times.

CONTENTS

CONTENTS

CONTENTS

PART 1

A 360° VIEW OF BUSINESS MODELS FOR WEARABLE TECHNOLOGIES

The smart device market has been dominated by a simple sales model—the selling of hardware devices themselves, including smartphones, tablets, PCs, cameras, media players and others. Before the internet, the exchange of money for goods was relatively straightforward: you pay for what you get. However, the internet has changed this pattern. We may use Commodity A for free, meanwhile we pay for Commodity B instead.

In the era of wearables, hardware devices will become secondary products. What we see today is the early stage of this industry with the selling of hardware device alone or hardware device plus a supporting app being the normal business model.

This is going to change. When the number of users accumulates to a certain level, the potential for complex analytics and use of data to extract value from the data flow will become the goal. This will then bring about profound changes in consumer behavior, transaction patterns, and business models. The sale of the device itself will not be the main profitable node, but rather the derived business that follows will be crucial to value creation.

CHAPTER 1

THE WEARABLE INDUSTRY IN CHINA TODAY

The Internet of Things (IoT) is gaining momentum and we are witnessing the start of an era that is more digitally connected than ever before. Wearables will be the carriers of the IoT and the market capacity will continue growing. The overall sales volume has increased from 84 million units in 2015 to 114 million units in the first half of 2016, a growth rate of 36%. Among global players, the Chinese wearables sector has impressed the world with rapid growth.

1.1 Smart Wristbands Dominate the Wearable Industry

A group of non-branded manufacturers in Shenzhen first launched various knock-off wearable devices. The market was soon transformed by the arrival of Xiaomi's Mi Band (activity tracker wristband), which soon gained tremendous popularity in China. The release of the Apple Watch in April 2015 was another milestone for the wearable technology market. In the same year, smartwatches for kids also appeared in the market and became the star product of the year.

Driven by these products, the wearable market in China soon grew to ship a million units each month. In June 2016, two million units were shipped in one month for the first time.

Volume and sales are both growing continuously. After the first half of 2016, the growth momentum was quite noticeable. The average price per device remained at RMB 300 (USD 42) in 2016. All four categories of wearables have seen similar trends of rising sales. Activity trackers are the most stable of all. Smartwatches for kids continue to grow in popularity. The future of this looks very promising and demand is growing robustly. In comparison, the growth of the smartwatch market overall appears to be slowing down, even falling occasionally. At the same time SIM-enabled smartwatches are increasingly popular, topping the other three categories in terms of growth.

The SIM-enabled smartwatch market in China is largely dominated by various domestic players, either small brands or non-branded. These low-priced products (RMB 170 or USD 27) are similar in design and many utilize all-purpose molds. The smartwatch market is largely dominated by international brands with an average price above RMB 1,000 (USD 140). Small players and non-brands are not competitive in this category. The market for kids' smartwatches, however, is dominated by domestic manufacturers. Internet companies have engaged in this category much more. Branded products have a similar market share to non-brands. The average price per item is around RMB 300 (USD 42). In the activity tracker market, both foreign and Chinese makers have their share, reflecting relative strengths. Compared to the pricier international brands, local Chinese brands are a better value and more popular among consumers. Established brands show better market position in this category and the average price per item is around RMB 150 (USD 21).

1.2   Changes in Market Price for Wearables

Affected by some low-price and non-branded products, particularly SIM-enabled smartwatches and smartwatches for kids, wearables on the Chinese market today have an average price of around RMB 300 (USD 42).

The performance of different price ranges shows that the RMB 0–100 (USD 0–14) price range took the highest share in the first half of 2015 due to the popularity of Mi Bands. The arrival of smartwatches for kids in July 2015 soon had many contenders with various products on offer, which is reflected in the sudden rise in the RMB 200–500 (USD 28–70) price range. In June 2016, two price ranges saw an increase in sales. Lifesense Band and the non-brands both drove the RMB 100–200 (USD 14–28) price range in this category, while high-end product sales pushed up the over RMB 2,000 (USD 282) market share as well.

1.3   Potential Breakthroughs in Product Function

Most wearables on the market today have similar functions. Step counting and heart rate monitoring are the most basic of these. 80% of all wearables on the market have a pedometer, mostly seen in smart wristbands and children's watches. Heart rate monitoring is largely seen in smart wristbands. Driven by Xiaomi's new release, the market share of products with a heart rate monitor is increasing too, reaching 30% of total sales.

We can examine first the functions offered by category of wearable device. Many SIM-enabled smartwatches have cameras, but few have health functions such as a heart rate monitor or pedometer. Lower-priced smartwatches have fewer functions. A NFC (Near Field Communication) function is largely available only for products over RMB 200 (USD 28). Most smartwatches for kids have the ability to make or receive phone calls, GPS, and pedometer functions. Few of these have heart rate monitors or other health functions, or NFC. Most activity trackers have pedometers and some can monitor heart rates. Few of these have NFC and cameras are only seen in SIM-enabled ones.

The existing wearables at this stage still have issues in terms of functionality. They tend to be homogeneous in function and lacking in creativity. They often have problems in terms of accuracy as well. No major breakthrough is foreseeable in these aspects, and consumer pinch-points are often missed. The entire market is therefore lacking rigid demand; therefore sales growth is slow.

However, the rise of the Internet of Things is an undeniable trend. Driven by that, makers of wearable technology should think carefully about how they should position their products, highlight functionality, and connect effectively with the Internet of Things and Big Data. Working these issues out will be a crucial step forward.

1.4   Shift in Marketing Channels

In China, the online channel is dominant for shipping wearables, accounting for nearly 90% of overall shipments. Major price differences are noted between online and offline channels, and the gap is widening. For example, the average online price increased from RMB 733 (USD 103) in July 2015 to RMB 1,199 (USD 169) in June 2016, and the annual average price in 2016 was around RMB 1200. On the other hand, the offline selling price dropped from RMB 293 (USD 41) in July 2015 to RMB 257 (USD 36) in June 2016.

Seasonal sales exhibit similar patterns between online and offline channels, but there was a marked alteration of peaks near Chinese New Year. Between July 2015 and June 2016, the average monthly shipments from online sales was 1.61 million units. The peak appeared in June, when 2.04 million units were shipped, and the lowest point in February, with only 930 thousand units sold. The maximum variance of the online market was 0.63 during this period. The average monthly shipments from offline sales were 150 thousand units. This peaked in January 2016 with 190 thousand units and the lowest point appeared in October 2015 at 110 thousand. The seasonal variation is very noticeable, particularly from the online channel.

Chinese buyers using different shopping channels also show preferences for different products. More buyers tend to use online channels for lower-priced items and offline for more expensive ones.

Low-end products priced below RMB 100 (USD 14) still had the highest market share online, reaching 40%. However, the popularity of Lifesense Bands pushed up the RMB 100–200 segment from 13% in Q2 2015 to 37% in Q2 2016, the only price range that rose year-on-year in Q2.

On the other hand, the offline market went in the opposite direction. Premium products pushed the market share of low-priced items down further. The RMB 500–1,000 (USD 70–140) range reflects the impact of XiaoTianCai (a smartwatch for kids), which rose to 36% as the leading market segment. In second place was the RMB 2,000+ (USD 282+) range, largely led by the increasing sales of Apple and Huawei products.

These price segments use different channel strategies. Most products in the RMB 0–500 range use solely online channels, while those in the RMB 500–2,000 range use a wider variety of channels. It is noticeable that there was more use of offline channels for more expensive items in Q2 2016 compared to Q2 2015, with the RMB 2,000+ range taking over to become the largest segment at 54%.

1.5   The Future of Wearables

Further segmentation and specialization is the likely future of the wearable market.

SIM-enabled smartwatches: Consumers have begun to shop more rationally for SIM-enabled smartwatches. This product category seems to have lost much of its appeal due to Samsung or Apple copies and the use of all-purpose molds. The early traction gained by non-branded products is gradually waning. Opportunities still exist for major brands, but the small players and non-brands started the price war too early, leading to severe price competition in this category. The entire production chain is developing and the demand for a telephone function in the device is becoming important. This market segment may experience some expansion in future with the potential entrance of established brands.

Smartwatches: Most wearable manufacturers maintain a strategic presence in this market segment. This is the major battlefield for both international and local Chinese brands. Competition in this segment will intensify. Product differentiation and strategic channel selection is crucial.

Smartwatches for kids: Established brands and non-brands are the major players in this segment as smaller businesses are gradually forced out. This market is moving towards two ends: premium brand products and low-priced non-brands. Branded products are fighting relentlessly for market share, by having more presence in different price ranges for wider consumer coverage. Manufacturers should focus on marketing and sales channels, looking at more traditional channels to reflect the nature of these products.

Activity Trackers: Xiaomi dominates this market segment in China. Several international brands have experienced poor sales due to comparatively higher price tags. Only a handful of local Chinese players see their sales rising. Most of these players focus strategically on specializing in sports and health. With successful marketing and promotions, they have managed to hold their market share while facing strong competition from Xiaomi. The highlights of the activity tracker market in 2016 are the advance of sports and health functions, as well as increased connectivity with different screens.

It seems that in this market segment, some products will focus on technology, product function, and performance, developing technical innovations on an ongoing basis. Some other products will remain light on technology, focusing on battery life and appearance instead. The future of activity trackers undoubtedly lies in the Big Data generated from users, in order to provide more precise and targeted services in return. Further segmentation in this market will continue, and we will see the coverage extend further into healthcare and extreme sports.

1.6 Consumers' Expectations for Wearables

Consumers expect the following from wearable devices:

1. **Appearance:** Consumers demand their wearables to look good. The devices are expected to look cool and fashionable.
2. **Battery Life:** A minimum of three days.
3. **Functionality:** As many functions as possible are preferred.

4. **Specialization:** Based on different market segments, the products need to be specialized in certain areas, and able to provide lifestyle advice.
5. **Smart Device Integration:** They must be able to connect with other products, to make life easier and smarter.
6. **Innovation:** Consumers are always interested in innovative functions. They are willing to try a brand that offers unique and creative functions.

CHAPTER 2

SALES OF HARDWARE AND DERIVATIVES

2.1   Market Outlook of Wearable Devices

Wearable technology has arguably been around for centuries—from the first spectacles that were made in Italy around 1290 AD, to watches dating to the 16th century and the modern wrist watch. Smart wearable technology builds on these innovations, combining advances in information technology. The trend for modern wearables can be dated to the 1960s and MIT's Edward O. Thorp. Professor Thorp noted in his gambling tutorial book *Beat the Dealer* that he had come up with an idea for a wearable computer in order to increase the odds of winning in roulette. He completed the device in 1960–1961 after collaborating with another developer. In June 1961, this device successfully improved the odds of winning roulette by 44%.

The first Fitbit tracker was launched in the US in 2009. In 2012, Google Glass kicked off the concept of wearable technology, soon followed by an explosion of various wearables in 2013. Some successful products started to appear in the market. Smart bands, or activity trackers, were the first type of wearables that won popularity. Among them are Fitbit, FuelBand, JawboneUP and more. These all focus on tracking human health data. With relatively simple technology and a ready-to-go supply chain, they soon gathered traction in the market. International IT giants soon followed suit. Apple, Google, Microsoft, Intel, Samsung and others quickly launched their own wearables. The product range became more varied, providing an ongoing theme for futuristic technology.

According to a forecast by IHS, the total global sales of wearable devices could reach USD 33.6 billion, with CAGR (compound annual growth rate) up to 22.9%. Obviously, future products will require more advanced technological capabilities than existing products. Such advances are expected in both software and hardware, as well as the capacity for industrial integration of the end product manufacturers. Wearable technology will rapidly improve consumer electronics. It is a common belief in the market that the growth of wearable devices will be even higher than smart phones and tablets, based on the industry's development to date.

Judging from overall shipments, sales of wearable devices were expected to increase from 90 million units in 2014 to 140 million units in 2015, with annual growth rates reaching as high as 62%. Smartwatches will remain the leader among all wearable devices, with an annual sales growth of up to 235%. The share of smartwatches in overall sales will gradually increase as well. Meanwhile, the market will see a growing diversity of wearable device offerings.

According to a report by Morgan Stanley, shipments of wearables in 2020 could reach 1 billion units under the most optimistic scenario, much faster than smartphones and tablets at a similar stage of growth. As the concept of wearable technology settles, wearables have entered the next stage, focused on product engineering and function development.

On the other hand, as consumers become more knowledgeable about wearable devices, they develop higher expectations for products: more appealing designs, improved functionality, and more innovative features. In the field of healthcare, consumers are no longer happy with simple data collection and presentation. They want greater accuracy and an improved understanding of the analytics generated from their data. When global sales reach a certain scale, manufacturers of wearable devices need to build unique business models that fit their products, while constantly refining their products according to market demand. No business activity is sustainable without an established business model.

Although manufacturers are engaging in various trials and experiments in this regard, we have yet to see any particularly successful business models for the wearable industry that can be replicated at scale. Let us take a look at the existing business models across all four market segments.

2.2   Xiaomi Band and the Apple Watch

Market research firm IDC reported the total shipments of wearables was 11.4 million units in the first quarter of 2015. That represents a 200% year-on-year increase. Ramon Llamas, Research Manager of IDC's Wearables and Mobile Phones division, said that against the normal sales decrease after the holiday season in the first quarter, the wearables market remained strong in growth, which sends a very positive signal. Meanwhile, the demand from emerging markets is also rising. Manufacturers are working hard to meet the demands and seize the new opportunities.

As shipments continue to grow, many manufacturers are starting to see profits. However, most of those profits arise from the sale of hardware devices

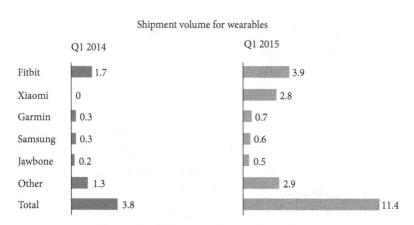

Figure 2-1   Shipments of wearables in Q1 2015

and their accessories. In the Q1 2015, total shipments of Xiaomi Bands reached 2.8 million units, accounting for 24.6% of market share, earning second place in the global market. Of course, most of the sales were in Mainland China. Some people attribute Xiaomi's success to its low-price strategy. A gadget selling at a price point of RMB 79 (USD 11) could encourage many people to try the product. But with this price, has Huami Technology, the manufacturer of the Xiaomi Band, made any profit at all?

We can examine some of the costs of producing a Xiaomi Band: the total cost of all components in a Xiaomi Band including the device body, wristband, motherboard, sensors, etc., is RMB 39.5 (USD 5.6). The cost after 17% VAT (value added tax) becomes RMB 46.2 (USD 6.5). Of course, R&D, shipping, and labor costs are not included here. Compared to the selling price of RMB 79 (USD 11), the gross margin still comes in at an impressive 100%. Shipments in the first quarter of 2016 were 2.8 million units, with total sales for the year forecast to exceed the 10 million mark, after taking the international market into consideration.

Xiaomi has not yet developed any value-added service to generate income beyond the selling of their devices. Even so, this shows how profitable this market is already. A well-marketed, quality product could be successful just on its own.

Apple provides another such example of this, when they introduced their long-awaited smartwatch, the Apple Watch, in March 2015. When Apple began the presale of the Apple Watch on April 24th, consumers from nine countries and regions placed their orders immediately. Dan Fromer, editor of Quartz Technology, noted that statistics showed around 1.5 million orders from the US were placed on the first day, of which 800,000 were within the first hour!

Of course, the price tag of the Apple Watch is what people care most about. Media reports say the starting price in the US is around USD 349. Putting exchange rate and VAT into consideration, it is around GBP 300 (USD 367) in the U.K. In China, the Apple Watch SPORT offered 10 versions with starting price of RMB 2,588 (USD 365), and the top of the range at RMB 2,988 (USD 421). There were 20 versions of the Apple Watch, starting from RMB 4,188 (USD 590) to RMB 8,288 (USD 1,167). There were eight versions of the Apple

Watch EDITION, from RMB 74,800 (USD 10,535) to the top of the range RMB 126,800 (USD 17,859) (Table 2-1).

Table 2-1 Apple Watch prices in China

China Mainland	SPORT	STANDARD	EDITION
38mm size	RMB 2,588 (USD 365)	RMB 4,188–7,888 (USD 590–1,111)	RMB 74,800–126,800 (USD 10,535–17,859)
42mm size	RMB 2,988 (USD 421)	RMB 4,588–8,288 (USD 646–1,167)	RMB 88,800–112,800 (USD 12,507–15,887)

Global Equities Research published a report noting that, since its launch in April 2015, orders of the Apple Watch reached 7 million units with total shipments of 2.5 million units. According to Slice, by mid-June 2015, Apple sold more than 2.79 million Apple Watches.

Apple not only sells smartwatches, but also launched a range of accessories. The Apple Watch's profit model to some extent is similar to a razor blade model, where razors are sold at lower prices for a higher margin in the sales of blades. Slice Intelligence has proven this through the analysis of emails.

Apple supplies many different types of straps for the Apple Watch, in a variety of colors and sizes. There are many options to choose from within this, in the price range between RMB 398 (USD 56) to RMB 3,588 (USD 505). The Sport strap is the basic option at the lowest price, at RMB 398, in two colors: black (42 mm) and pink (38 mm). The Milanese Loop, the Classic Buckle, and the Leather Loop straps are all sold at a price point of RMB 1,158 (USD 163). The Modern Buckle is RMB 1,928 (USD 272). The Link Bracelet, made of a stainless-steel alloy, sells at the highest price, RMB 3,588 (USD 505).

The actual manufacturing cost of the basic Sport strap, priced at USD 49, is only USD 2.05. According to statistics, in every 2 million people, over 20,000 have bought an Apple Watch strap. 17% of buyers bought more than one. Slice says this data is in line with the data from the US Department of Commerce and Amazon.com.

Judging from the iPhone experience, it is very likely that the Apple Watch could become a leader in smartwatches in the future. Cliff Raskind from Strategy Analytics predicted that total annual smartwatch shipments in 2015 could reach 28.1 million units, achieving an annual growth of 511%. Most of this growth would be from Apple Watch sales, with annual shipments of 15.4 million units. Raskind believes that the Apple Watch will soon become the top smartwatch in the world, taking up to 55% of the entire market share.

People are optimistic about the Apple Watch not only because it has many great features as a device, but also benefits from factors such as brand awareness, a customer loyalty, a strong and extensive retail network, as well as a comprehensive app ecosystem provided by Apple. All of these are guiding the future development of the Apple Watch.

2.3   Which Devices Are Profitable

The greatest value of wearable devices doesn't lie in the hardware, but rather in the massive volume of data generated from human bodies. Existing technologies and industrial development are not yet able to fully capitalize on this data, therefore early wearable manufacturers have focused mainly on selling the devices. This represents not only a quick win but also relative simplicity in product management.

How to make best-selling wearable products is therefore the focus of every player in this market.

1. Work on the design.
Each product must possess more than just looks and prove itself to be easy to use. It should be comfortable to wear, with fewer but more effective functions. With mobile internet, consumers are increasingly picky. They want more in the products. If a product fails to tick these boxes, it will be thrown to the back of a drawer once the novelty is gone.

This is an era of individualistic and differentiated consumers, making the design ever so important. Consumers are not only buying goods for their practical use. They are looking for fashionable looks and something unique. The products they buy need to reflect the taste of their owners and provide sensory and emotional satisfaction. With multiple potential formats available, wearables provide great opportunities for creativity. When design is well done, products will certainly gain consumer attention.

2. Build killer applications and functions.
It is in fact rather difficult to discover genuine customer needs through market research. If you ask people what they want from their smart devices, the answers are often obscure. They may offer some suggestions, but the true needs could be deeply hidden.

The iPhone has taught us well in this regard. Before the birth of smartphones, users had never asked for a phone with such functions. Steve Jobs first discovered these potential consumer needs and created an iPhone accordingly. It has changed the whole world in terms of communication, socializing, or even our way of life.

The actual needs of users are often hidden behind some superficial demands. Only with insightful research and analysis could they be discovered and turned into tangible functions. If a wearable device wants to gain customer recognition quickly, the R&D behind so-called "killer" applications and functions is the key. Nothing else works without this. Meanwhile, killer apps and functions also provide a technological barrier for self-protection against homogenization, allowing a smart device to protect its place in the market for longer.

3. Make bespoke products and provide a customized service.
We are seeing a return to user-led design. Even industry leaders like Google Glass cannot rely on its technological edge to ignore consumer needs. They respond to them by introducing different glass frames and colors, to cater for customer preferences and to maximize market share. Apple has also launched different models with different price ranges based on consumer segments. Everyone likes to be different in some way. A generic product would be difficult to sell when variety is so prevalent.

4. Win with soft values.

When the Apple Watch was first introduced, many people could not accept the price. Did Apple worry about people being put off by their high prices? It could be argued that this was a deliberate move. What enables Apple to capture the hearts of so many? It is not only because of its hard values but also its soft values. The exterior design of Apple products and good customer experience from an easy and hassle-free system are the main reasons customers cite for their attraction to the brand. In the mobile phone industry, the iPhone is always relatively expensive, nevertheless endless consumers are content to pay for them, even if it is really beyond their means.

5. Build wearables into luxurious items.

The launch of the Apple Watch has made many traditional wristwatch makers turn their focus to digitalizing their products. Today, smart devices can be considered fashion in and of themselves. If wearable technology can leverage this momentum and step into the luxury goods market, by adding value to products through brand effect, premium material, and fine handicraft, it will surely increase product sales revenue. For example, the sales price of the Apple Watch Edition ranged from RMB 74,800–126,800 (USD 10,535–17,860). Equipped with an 18K gold watch case, this version was visually distinct from the other two versions. The price difference is largely determined by the choice of strap, from the cheapest fluoroelastomer straps to the most expensive modern-style leather straps. This price range obviously is already at the level of luxury wristwatches, which likely still appeals to many customers. When fine workmanship is combined with artificial intelligence, it turns out to be the most "in" item of the moment. For affluent and trendy people, what could be more attractive?

Clearly, the business model of selling hardware alone cannot bring long-term profitability to companies, nor can it maximize the return value, as the most profitable model of wearable technology is based on extracting and implementing Big Data. When this model is up and running, the devices themselves will no longer be the main source of profit, rather they would change to become "accessories."

CHAPTER 3

BIG DATA, THE NEW FRONTIER

The analysis of Big Data, including data from wearables and other data sources, is still in its infancy. Many companies are innovating in this area. Exploitation of Big Data offers great potential for companies producing wearables in terms of value generation. There are early success stories of companies that have been early adopters, some of which are highlighted in this chapter. A true Big Data service will not only bring direct return for business, but also tremendous convenience in daily life for consumers.

3.1 When Advertising Meets Big Data

Big Data has reinvented the advertising industry overnight by enabling highly effective and precise marketing.

Businesses traditionally spend a great deal on gaining market intelligence. Whether it is knowing who their customers are, what triggers them to buy, where and when they are likely to buy, at what price range, where they might use the products—the list is endless. With the rise of wearables, we are moving towards a digitalized horizon for all consumer behavior. Businesses and marketing consultants can perform data analytics to determine the most fact-based and accurate marketing ideas. Advertisers can therefore impact the target audience most effectively.

To best capture the advantages offered by Big Data, business models framed around collaboration with social media companies or platforms offer great potential. Potential collaborators include Google, Facebook, Baidu, Tencent, and others. Why is this the best? It's simply because these companies have control over all the information that consumers generate while searching the web, networking, chatting and emailing. Baidu's search engine index reveals how topical a specific keyword is at the moment and demographic characteristics of those looking for certain keywords. For instance, the search results of the word "diet" could reveal the type of diet people are more interested in. This information is of great potential value to companies in this business, though it may seem worthless to the individuals searching. By analyzing this information, overall customer preference can be understood. Products can therefore be developed targeting specific customer segments.

Other personal information—including daily interests, food and clothing preferences, and location—used to be hard to gather at a large scale before the arrival of Big Data. Now Facebook or WeChat provide an endless source of data through users' online chatting, sharing, and blogging activities.

The greatest value of search engine companies like Google or Baidu lies in the fact that they retain user search histories on which the logic processing of Big Data service is based. Gmail publishes targeted advertisements based on content scanning of the user's email. People sometimes joke about receiving advice on how to commit suicide after using the word "suicide" in business emails. In some ways, this could be seen as a violation of privacy when advertising is targeted in this manner. And yet precise marketing also helps meet user needs far better than ever before.

In 2013, Procter & Gamble launched a campaign in China under the name of "Pretty Mom," using search engine data from Baidu. This campaign was awarded the Best Digital Marketing Case of the Year. They had a clear customer group to target: young mothers who were not using disposable diapers. It worked.

Baidu is the number one search engine in China with the best Cloud system as well. It has established the most powerful Big Data platform in China through its talent-rich Big Data Unit. As Baidu adopts an open platform strategy, it is also the preferred choice for more businesses and brands.

What Baidu offered Procter & Gamble was a convincing statistics-based argument: mothers using disposable diapers have on average 37 minutes of additional free time per day compared to those who don't use them. Baidu then further analyzed the activities of mothers during that additional free time and discovered that apart from the expected childcare concerns, they also tended to use the free time to focus on body recovery and pampering themselves after childbirth. This is how Procter & Gamble chose the theme of "Pretty Mom" to launch their online campaign. Sales of Pampers soared in the following season as a result.

3.2   Build Data Analysis Models

Using apps or other means, branded wearables have all adopted Cloud technology to store and exchange data. Analyzing this data could help generate new business models. What wearables provide is a channel for sequential and continuous data collection. Heart rate, blood pressure, and other information are uploaded onto the Cloud from the wearables and synchronized with mobile apps while remaining stored on the Cloud.

The collaboration model between wearables and Big Data is integrated into the configuration of the Big Data platform and is needed for making use of preliminary data. Specific measurements of the human body from activity trackers are not of much importance to ordinary users, but they are the inputs for later data processing. Data analysis companies specialized in processing such data have become an emerging trade.

When they set up data models that can compare and analyze input data, visual results could be generated instantly. Companies are able to share their results with healthcare organizations, commercial firms, and governmental agencies, or even insurance companies. Of course, medical professionals need to be involved to advise on health-related decisions. But with that information, healthcare organizations would be empowered to provide health alerts or preventative services to their customers. Insurance companies could adjust premiums accordingly. A complete production chain is therefore viable from the starting point of wearables.

Based on research on business models of wearable technology in the US, there are insurance companies in America that are already incorporating wearables into their businesses, sometimes building unique business models. We can mainly see two types of models.

In the first type, insurance companies pay wearable producers when their customers use those services, while in the other type, insurance companies encourage healthier lifestyle habits by adjusting premiums according to customer activity records.

The diabetes management company WellDoc is an example of the first type of business. Currently they are using a "mobile + Cloud" diabetes management platform, focusing mainly on the mobile healthcare side. But greater value would be possible if wearables were involved in this model.

Currently, WellDoc relies on mobile phones to record and store user blood sugar data to be uploaded to the Cloud. After analysis, it provides personalized feedback to the patient and reminders to doctors and nurses. This system has proven to be clinically effective, it is economically viable, and it has been approved by the US Food and Drug Administration (FDA).

The BlueStar app by WellDoc provides timely messaging, guidance, and education services for patients diagnosed with Type II diabetes who are in need of regular drug treatment.

WellDoc charges fees according to the services it provides to users. As this service has been proven to be effective in reducing long-term costs for healthcare insurance companies, so far there are two insurance companies paying over USD 100 per month for clients with diabetes who use this "Diabetes Management System."

The other type of business model is based largely on data mining and data application. Wearables are able to monitor and record real-time body statistics through various embedded sensors. Insurance companies could use such data to know their customers in terms of lifestyle and daily activities. To reward healthier lifestyles and exercise, they could lower premiums as a result and increase premiums for those who fail to achieve certain targets.

This is mutually beneficial for insurers and for those who adopt healthier lifestyles. Insurance companies save costs on paying out claims, and customers have more incentive to live healthily. The healthier the customers, the lower the costs.

In the US, most medical insurances are paid jointly by employers and employees. A wearable-associated health insurance scheme could reduce such costs for employers and encourage employees to keep fit and live healthily. This in turn increases workforce productivity. Wearables thus offer many potential returns, killing multiple birds with one stone.

CHAPTER 4

SYSTEM PLATFORM AND APPLICATION DEVELOPMENT

As one of the earliest entrants in the activity tracker market, Fitbit is one of few companies that has managed to take both wearables and their health tracking app into the mainstream market.

The mobile app from Fitbit can be used together with a Fitbit tracker or a smart scale, to encourage users to live more actively and eat healthier. This app allows Fitbit's trackers to synchronize with either Android or iOS devices.

The Fitbit trackers can record steps, distance, calories burned, and other data automatically and synchronize the data onto the mobile app without any manual operation from the users. In the "track exercise" module, the user only needs to open the module and press the "start" button on the app or wearable device to track activity and movement. Automatic tracking has also become prevalent among these. Users can also set up voice reminders or link with music to better pace their exercise routine.

There is also the "sleep analysis" function. Sleep can either be recorded automatically by the tracker, or manually added by the user.

The app also includes a function to record water intake. Again, users can add this information manually or use the quick adding function. However, this is not the most useful or accurate function. Most people find it hard to recall exactly how much water they have consumed throughout the day.

Fitbit has performed really well since its appearance on the horizon. Half of the appeal of its trackers comes from the mobile app. Without the app, the functionality of the trackers would be much diminished. In the future, wearables will be an entirely different type of device, independent from smartphones.

The existing wearable + mobile app formula in essence only uses the mobile device as a display screen for the wearables. When virtual reality technology develops further for commercial purposes, the smartphone might lose its place in this business model and could well be replaced entirely by wearables. Or maybe they will survive in a different format. They may turn into what might be called "wearaphones," a product among many other wearables to arrive in the future.

4.1   Major International Big Data Cloud Service Platforms

The wearable technology industry so far has been rather fragmented, reflecting the reality that the current Big Data Cloud Service platforms are not yet well integrated. Many wearable makers only produce and market the devices or stop at the device + mobile app stage. Many companies, especially startups, do not have the capacity to build a comprehensive platform for Big Data analytics.

International IT giants however are all working on their own Big Data platforms, hatching plans to conquer individual markets. These include Google Fit, Apple Health, Microsoft's HealthVault, and Samsung's Digital Health and SAMI Cloud platform.

It's clear that many major players are working to build mobile healthcare platforms to attract new customers. By doing so, IT giants are making inroads into the mobile healthcare market through wearable devices. By leveraging their own platforms, these giants are hoping to gather more user health data from third-party device and application providers in order to work with clinics, hospitals, and other medical service providers. The goal is enriched data sharing and the eventual creation of a healthcare platform to better serve the needs of all users.

1. Google Fit

Google entered the e-health market in 2008 with the launch of Google Health, a health data sharing service that partnered with CVS Pharmacy and companies like Withings. Users could create a personal data profile on the platform in order to receive healthcare services more conveniently. Google Health failed to incorporate mainstream medical services and insurance companies into the system, and it was shelved in 2011.

Nevertheless, Google is undoubtedly a very popular self-diagnosis platform, thanks to the powerful data flow from the search engine. According to Pew Research Center, 35% of Americans prefer to self-diagnose symptoms using Google.

Google obviously will not abandon the e-health market. During the Google I/O 2014 conference they launched Google Fit, focusing on tracking and sharing user activities. In August 2014, Google released the Google Fit Preview SDK to developers. As with Apple, Google Fit also provides an activity tracking data platform for third party apps and API for data storage. In other words, Google Fit abstracts data from third party activity tracking apps to generate Big Data. Its overall strategic direction so far is still focused more on individual users' stats, not support for general medical institutions. If Google Fit is one day connected with Google Glass and other wearables, its potential value in the healthcare industry would put it in a very strong position.

2. Microsoft HealthVault & Microsoft Health

Microsoft HealthVault was established in 2007. On this platform, with one registered account, a user can set up and maintain their health record. It works like an information safe. It has open interfaces to connect to device makers or insurance companies, but it is up to users to decide who they want to provide that information to.

Users can upload monitored data from other devices to HealthVault, to be analyzed in order to offer users recommendations of potential solutions.

The Next Web (TNW) interviewed Steve Nolan, head of Microsoft HealthVault, by email to discuss the future prospects of digitalizing medical records. Nolan expressed his hope to turn HealthVault into a brand like PayPal or Visa, offering crucial information to health care providers to facilitate optimal care and treatment for patients.

While other IT giants focus on releasing their own health management platforms to get ahead of this wave of wearable technology, Microsoft on the other hand has quietly upgraded and perfected its own platform, developing a sizable user base ready for launching its own wearables in the future. Customers can upload fitness data including blood pressure, breathing rate, blood sugar and more from 233 different wearables, and exchange data with over 160 apps. Although HealthVault is a practical service, it has not been well received. This may be because of its poor customer experience. Its underlying programming has limited compatibility and users have limited understanding of the platform's full functionality.

In 2014, Microsoft launched another health data platform called Microsoft Health, which offers a Cloud storage service to consumers and businesses. Microsoft Health is able to integrate data extracted from various health and fitness devices stored in the Cloud. Users are also able to compare data from various devices with data stored in the Cloud. By comparing steps, burned calories and heart rates using Microsoft's "Intelligence Engine," valable conclusions can be made, such as which type of exercise will be most effective.

Microsoft explains how this could provide useful recommendations on daily exercise routines and lifestyle patterns. By integrating activity tracking data into the Microsoft Office calendar, emails, and GPS locations, Microsoft Health Intelligence Engine could help assess a number of issues necessary for living a healthy life – from the best way to fit gym time into work schedules, eating healthy meals, whether too many meetings affect sleep patterns and so on.

Collaborations have been established between Microsoft Health and multiple device suppliers or service providers including Jawbone Up, MapMyFitness,

MyFitnessPal and RunKeeper. The next step is to provide users with the option to connect Microsoft Health with HealthVault to share data with healthcare providers.

3. Apple HealthKit and ResearchKit
Apple released the mobile app HealthKit during its annual developer conference on June 2, 2014. This app is empowered to integrate fitness data collected from Apple devices—including the iPhone, iPad or Apple Watch — for analysis.

When Apple released iOS 8, the built-in HealthKit platform won support from over 900 third-party fitness apps and activity trackers, including Jawbone Up, MyFitness Pal, Withings Health Mate and others.

Apple has also reached agreements with many established US hospitals, including Duke Medicine, Ochsner Health System, and the Mayo Clinic, to name a few. Cedars-Sinai Medical Center recently updated its electronic patient record system and adopted the HealthKit system to run over 80,000 copies of patient data. This was one of the largest in scale at the time.

What HealthKit is trying to achieve is an integration of fitness data from different devices and apps, and to incorporate that integrated data with electronic medical records for review by doctors or for real-time tracking and alert functions. This will surely be extremely useful.

On March 10, 2015, Apple COO Jeff Williams announced the release of another new healthcare app, ResearchKit. One unique aspect of this platform is that it was open source. By collecting user data, ResearchKit could work together with medical institutions and specialists on various studies.

Jeff Williams also expounded on the motives behind ResearchKit, that is, the facilitation of medical research. There are many constraints on medical research today. Volunteers are hard to recruit for studies. Data accuracy is an issue, as subjective data from traditional paper questionnaires is not satisfactory.

Apple is aiming to enlist the planet's millions of working iPhones to record user health data and use this data for medical research.

In terms of privacy protection, Jeff Williams explained that users would be informed of the related risks when participating in a ResearchKit program, and would be asked if they were willing to share personal information with researchers and their collaborators. Therefore, the users had the final say on access to their data.

ResearchKit partnered with quite a few institutions and developed five apps, including the Parkinson mPower Study app for Parkinson's Disease, GlucoSuccess for diabetes, MyHeartCounts for cardiovascular diseases, Asthma Health for asthma and Share the Journey for breast cancer patients.

All these apps quantitatively track and monitor users' body conditions and perform analysis accordingly. For example, the app GlucoSuccess, a collaboration between ResearchKit and Massachusetts General Hospital, reminded users to follow five steps on a daily basis, notably:

1. Measuring body weight
2. Tracking activity (using iPhone as a tracker with app open during daily fitness)
3. Answering two questions, about quality of sleep and self-assessed foot health
4. Monitoring of food intake (using the free app "Lose it!")
5. Monitoring glycated hemoglobin (HbA1c) test results

According to press reports, Apple had planned to collect DNA test results through the ResearchKit app. Such data was collected mainly by research partners and stored in the Cloud, and was permitted for use in medical research and studies. For example, the University of California is collecting DNA information from pregnant women to studying the causes of premature delivery.

There are major differences in the user base and product positioning between ResearchKit and HealthKit. ResearchKit focused mainly on medical research, while HealthKit aimed to be a personal fitness advisor for individuals. In terms of the number of apps, HealthKit had many more than ResearchKit, and data on the HealthKit platform was not used for medical research. However, ResearchKit was used to better leverage the strengths of HealthKit.

The medical industry will surely evolve alongside developments in these areas. The cycle starts from user health data integration through the health management platform, followed by research and analysis using ResearchKit. When precise and effective results and recommendations are derived, they are returned to medical specialists and users and the loop is complete.

4. Samsung Digital Health

On November 13, 2014, Samsung launched a platform remarkably similar to HealthKit during its developer conference in San Francisco. This healthcare solution platform is named "Digital Health." It gathers user data from mobile and wearable devices, then provides lifestyle recommendations for the users. This data is kept in the Cloud for user access at any time. Samsung has also partnered with Nike, Aetna, Stanford University and the University of California. Samsung believes fitness devices and apps will emerge as important preventative health measures in the long run.

4.2　The Major Big Data Cloud Service Platforms in China

Patients today have higher demands for medical services than ever before. People require better-managed and quickly accessible personal medical records, as well as more advanced and cheaper medical care. On the other hand, medical service providers are facing increasing pressure to lower costs, improve treatment results, increase efficiency, and expand services under ever-changing regulations.

The arguments over the Affordable Care Act were not limited to the circumstances in the US. In fact, it is a challenge across the globe, especially in China. People have an urgent need for better healthcare services.

During the National People's Congress and Chinese People's Political Consultative Congress in 2016, Premier Li Keqiang proposed a policy of accelerating the Chinese economy by leveraging the momentum of "Internet+." With mature internet technology, many industries are upgrading rapidly. We have seen many internet companies entering the healthcare industry in China.

Technological advances hold the key to solving many challenges. By eliminating redundancy, new technology and data-sharing models can enable safer sharing of medical information. A Smart Medical Cloud system could be the foundation of the entire smart medical care industry. An Electronic Medical Record (EMR) delivered through Cloud computing is one example of a technological advancement that would greatly improve patient care and reduce costs. It would not only be more accessible, flexible, and efficient, but also cheaper and more secure. The following section outlines some of the major players already engaged in this market in China.

1. Alibaba Health Cloud

Alibaba Cloud is the Cloud computing arm and business unit of Alibaba Group. It was founded in 2009, and in 2010 Alibaba Cloud's first data center officially opened. Through Alibaba Cloud, users can access large-volume remote computing, storage, and Big Data processing services.

Alibaba Health Cloud is an internet healthcare solution under the umbrella of Alibaba Cloud. On April 1, 2015, the first Alibaba Cloud Hospital was launched. This cloud platform is called Butterfly Valley Medical, offering medical system integration, and a network platform for a range of medical services. Business areas include doctors, medical institutions, patients, medical insurance, health management, pharmacies, regulators and more. It aims to build a tiered healthcare management structure and a referral system among hospitals.

Wang Yaqin, CEO of Alibaba Health, explained, "Butterfly Valley Medical is not a traditional hospital concept, but a platform set up for doctors. Doctors are equal partners, and achieve a win-win collaborative result. Doctors' services are reviewed by patients on the platform, building up their reputation and worth. On the other hand, the platform also works with physical entities to support the mobility of doctors across different locations, hoping to grow the industry online and offline together."

Wang expects this platform to create opportunities for multiple practicing locations. "The Cloud Hospital platform aims to facilitate the building of doctors' individual platforms, to enable healthcare professionals to be flexible and not bound to a specific medical organization."

Healthcare is a crucial part of Alibaba's corporate vision. The integration of the healthcare business and Cloud computing will become one of the key battlegrounds in the strategic positioning of Alibaba among its competitors for the internet healthcare business of tomorrow.

2. Tencent Health Cloud

Tencent Cloud was initially developed for online gaming. After years of development and service expansion, it has become a strong arm of Tencent. Tencent Cloud features an open ecosystem to connect end user capacity for a collaborative Cloud ecosystem, linking business partners, customers, and end users together.

Tencent has become the social media leader in China with its WeChat software. By leveraging WeChat, Tencent made its first step into the Smart Healthcare arena through WeChat Hospital Booking. So far, over 1,200 hospitals employ some "Smart Healthcare" through the WeChat platform. Nearly 100 hospitals use the WeChat platform throughout the process of a hospital visit and over 120 hospitals support WeChat Hospital Booking. In total, over 3 million patients have benefitted from this service. WeChat Smart Hospital was launched in 2014. It is based on "WeChat Public Accounts + WeChat Pay," which interfaces with mobile businesses. It is able to verify a user's ID, perform data analysis, settle payments, manage customer relationships, offer after sales service, and protect consumer rights. By optimizing connectivity between doctors, hospitals, patients and medical devices, the entire treatment process is simplified.

3. Baidu Health Cloud

On July 23, 2014, the Beijing municipal government, Baidu, smart device suppliers, and service providers jointly released the "Beijing Health Cloud" platform. According to Hou Zhenyu, Chief Software Architect of Baidu Cloud, there are three levels in the structure of Beijing Health Cloud: sensor devices, the Health Cloud platform, and healthcare services. Hou explained that the three levels reflect a progressive relationship from bottom to top, completing the cycle that tracks users' health conditions. So far, eight devices have been connected to the system including an activity tracker, blood pressure monitor, ECG, body weight scale, and body fat analyzer. Some of these are under Baidu's own smart device brand, Dulife.

Users track their own fitness data through these devices and upload the data to the Cloud. After Big Data analytics, users are then offered fitness and diet suggestions, health management advice, and a remote heart monitoring service along with many other health services.

Baidu Vice President Li Yuanming explained in an interview, "In this age of 'mobile internet + Cloud computing + Big Data,' the direction for innovative technology is clearly "software + hardware' and 'online + offline.'" This is indeed what Baidu is doing. Bringing together wearables, Cloud computing, Big Data processing capacity, and professional services, Baidu has constructed a strong foundation for Baidu's Health Cloud platform, on which each user can set up a digital health profile for free in return for overall health management services.

Baidu's biggest strength is undoubtedly its Big Data resources. As an established search engine company in China, Baidu has accumulated a colossal amount of data, which facilitates all sorts of potential data applications. For example: epidemic surveillance, disease prevention and control, clinical research, medical diagnosis decisions, coordination of medical resources, and family healthcare at a distance, to name a few. The target for Baidu in the healthcare industry is to achieve the true "connection between people and services" through the internet.

4. JD Health Cloud

JD Health Cloud provides users with services like health data recording, a personal health profile, and personalized healthcare advice. As an important part of JD Smart Cloud, JD Health Cloud gives its users access to comprehensive personal health services.

Devices that can connect to JD Health Cloud could be used in many aspects of daily life. An activity tracker automatically records data from the morning exercise routine, which contributes to a personalized fitness and diet plan delivered through smart data analysis. A healthy menu is suggested at breakfast time, the consolidated result of recent activity and body condition. A vibrating alert is set to remind a desk worker to stretch their legs regularly. When the user returns home at the end of the day, standing on a smart scale records ten measurements which are uploaded to the Cloud. The Health Cloud keeps on monitoring the user at night, tracking sleep quality and offering suggestions based on long-term data.

Wearable device makers can also store user fitness data on JD Health Cloud by connecting those devices with the platform. Using that data, third-party providers could then offer healthcare, fitness, and wellness services to the users.

JD Health Cloud has been working with dozens of wearable device suppliers in China and overseas, including Connaught Labs, Codoon, T9, Dnurse Technology, HiWatch, Goccia, and Sanyuki. Together they provide services for many aspects of users' health, such as exercise, sleep patterns, blood pressure, blood sugar, body composition, heart rate, and movement.

5. *Chunyu Health Cloud*

Founded in 2011, Chunyu Mobile Health is the developer of the Chunyu Doctor app, offering virtual health consultations, appointment booking, and many other functions. It is a mobile healthcare app for self-diagnosis and free consultations with doctors, providing solutions for users' medical concerns.

Chunyu is working on its Electronic Health Record (EHR) system for filing users' health data. Zhang Yue, CEO of Chunyu Mobile, summarized the challenges as follows: "EHR can be seen as Big Data for healthcare. Real-time health measurements, historic medical records, physical examinations and DNA test results, and healthcare consumption behavior are the four major parts forming the entire data pool." He believes that EHR will be the operating system for doctor-patient communication in the future. It enables doctors anywhere on the planet to provide effective intervention and guidance to meet patients' medical needs.

The basic logic and operation model of Chunyu's EHR is based on collaboration between Chunyu doctors and various solution providers, to jointly offer service packages to users. For example, the first collaboration was with ETcomm, a mobile health monitoring app and technological solution provider. After ETcomm connected with Chunyu data, doctors on the Chunyu platform were able to retrieve reference data from ETcomm devices, including blood pressure, blood sugar, ECG readings, and other useful information.

Data collection is done through hardware devices, and further data cleaning and processing is completed by Chunyu's medical teams. The aim is data standardization through user segmentation based on different chronic diseases (Table 4-1).

Table 4-1 Data collection by user segment in Chunyu's EHR

Hypertensive patients – Cardiologists, GPs	Diabetes patients – Endocrinologists, GPs	Heart disease patients – Cardiologists, GPs	Female patients – Obstetricians and Gynecologists, Nutritionists	Child patients – Pediatricians, Nutritionists
+ Sports	+ Sports	+ Sports	+ Sports	+ Sports
+ Body weight	+ Body weight	+ Body weight	+ Body weight	+ Body weight
+ Body fat	+ Body fat	+ Body fat	+ Body fat	+ Body temperature
+ Blood pressure	+ Blood pressure	+ Blood pressure	+ Sleep	+ Sleep
+ Heart rate	+ Blood sugar	+ Heart rate		
	+ Blood oxygen	+ Blood oxygen		
		+ ECG		

Zhang Rui, founder of Chunyu Doctor, explains that the data sorting process is so far largely based on the needs of the doctors. When specialists are consulted by certain types of patients, they would like to see patient records arranged in the same standardized manner, fixed and non-changeable. Doctors benefit from such data processing through time saved and reduced testing costs. The back office could make use of previously unavailable long-term tracking data to produce charts and tables for doctors' reference. When any abnormal condition is detected, it could alert doctors to enable an immediate response.

CHAPTER 5

THE OVERLAYING OF
BUSINESS MODELS

Wearables are indeed a special type of product. They represent crossovers between traditional consumer electronics and the emerging technology industry, meeting consumers' personalized needs. They offer consumers new functionality to facilitate greater health and convenience in their lives. However, wearable suppliers face challenges in the task of building sustainable businesses in the longer term.

Although design has been seen as the key to the success of particular wearables, without a proper business model sales may not take off as desired. Among many potential business models, an overlaying of different models could be adopted as the main business format for the future leaders of the wearables market.

Let us take Google for example. Google owns its own brand of wearable devices, Google Glass, and the Wear OS platform. Likewise, Apple has the Apple Watch and apps to use together with it. In the healthcare industry, Apple also has HealthKit and ResearchKit.

We see Google was successful using the standalone device model with its marketing of Google Glass. But they also expanded into many different business models, leveraging Big Data, adding advertising to Google Glass, and collaborating with other businesses for training and engineering projects. Apple's smartwatch mainly focuses on personal fitness. Apart from the device + accessory sales model, they also support users with two health platforms. Using data collected from users for medical research is another model that Apple is working on.

5.1   Building a Stronger Internet of Things by Adopting Multiple Business Models

If a company wants to succeed in the Internet of Things, it needs to take care of five key aspects: quality hardware, an independent system, application development, a Big Data Cloud service platform, and a social media platform.

The IT giants leading in this area have all managed to do well in the first three aspects. Examples include Apple, Google, Samsung and others. They are currently at the stage of building Big Data Cloud service platforms but have not yet seen major breakthroughs. What is happening to all of them? With their own data platforms, the value of the data has not yet been fully mined, and the raw data is not without error. This is a frustrating issue for many users. This represents a bottleneck in achieving the last aspect, the social media platform. Capitalizing on that abundance of data is thus difficult to achieve.

The ability to process, analyze, and generate feedback from data is one of the key elements contributing to user stickability. When that stickability is lost, users are very likely to give up entirely. On the other hand, a small and scattered circle on social media could also push users away to more popular sites. Online social gathering is often not the aim, but a natural result of wearable technology gaining popularity.

These IT giants, as early IoT explorers with both financial and human resources at hand, are full of ambition. What they have done so far has been to contribute to the development of this industry, but inevitably some will win and some will lose, as a result of the risks that early experiments entail. The real battle of the IoT age is not about devices alone, but rather about the killer weapons: the data and the platform. Just as with iOS versus Android, the battle of the operation systems is in fact the core competition splitting the smartphone makers into two groups, rather than the devices themselves.

Google has not been a major supplier of smartphones, but it contributed the Android system for many. In the IoT era, Google has gone straight into wearables, smart home devices, self-driving cars, and many other areas. The Android Wear platform they developed is the first platform released for smartwatches. It really shows how determined Google is to exploit all that the IoT has to offer. I believe the next move with system platforms will be to increase openness and integration. The self-developed systems from Apple, Google, Microsoft, and Samsung will eventually be able to connect and exchange data in the background.

IT giants benefit from their leading positions in this industry, as they are able to implement many different business models simultaneously. This is a major barrier for tech startups, who simply cannot afford the costs. We can see many startups in China and overseas using crowd funding to finance initial product development. Lack of investment funds forces these companies to focus on device sales to gain funds sooner. Most of these products are relatively simple in terms of technology and production chain. These intelligence-light products avoid the higher production costs of more complex technology. Product failure is something these companies cannot afford.

Therefore, startups generally adopt the business model of creating a popular product to market in return for quick wins and rapid returns on investment, or for further financing possibilities. Their contributions to the industry are limited to the sale of hardware devices. They improve the appearance and design details to make the devices more attractive and comfortable, or they upgrade parts of the devices to improve battery life or make data acquisition more accurate. These startups struggle to get involved in Big Data analysis or platform building. They simply do not have the capacity or infrastructure to support such activities. Instead, these companies are more likely to focus on device making alone, and seek to be connected to Big Data platforms provided by others, such as Microsoft HealthVault, an open platform for fitness apps to facilitate the upload and management of data. Or they could choose to use the data for research or analysis by collaborating with medical institutions, insurance companies, or other third parties.

5.2   The Great Opportunity of a Focused Model

As previously noted, the most profitable business model may be a combination of co-existing business models. However, this is not a strategy fit for all businesses. Most startups would not be able to adopt such a model. Industry leaders, on the other hand, are best suited to employ this strategy, as they have the necessary funds to fuel the complex structures and the human resources needed. It is still better for the start-ups to choose the standalone device model. They could follow this route and explore opportunities in the following areas:

1. Using market segmentation to build system platforms. System platforms can be built according to vertical market segmentation based on product types, including activity trackers, smart clothing, smart footwear, and more. Such platforms are less technically challenging, relatively, and easier to optimize around focused targets.
2. Building technologies for the production chain. Technologies used in nodes along the production chain can be used as entry points, such as voice interaction, batteries, smart materials, sensors, chips, and so on. Such technology-focused business models can create competitiveness, as long as the development team is equipped with sufficient technical resources and capacity.
3. Providing application-led solutions. Focusing on one portion of the production chain, these companies may look at providing apps, algorithms, technical solutions, or manufacturing for related technology startups, including designing crowd-funding schemes, with the aim of building unique strengths to provide solutions in a specific area.

As the entire wearable technology industry grows rapidly, technology and products are constantly upgraded. All parts of the production chain are gaining momentum and presenting even more opportunities, so choosing any entry point may be a great choice for aspiring entrepreneurs looking to start businesses with wearable technologies.

CHAPTER 6

POLICIES SUPPORTING THE DEVELOPMENT OF WEARABLES IN CHINA

The Chinese government has paid significant attention to the wearable industry in recent years. The National Development and Reform Commission published notices demanding more focus on mobile internet applications and the development of various commercial-scale wearables, particularly low power consumption wearable systems, new models of human-computer interaction technology, and new sensor technology for wearables. The government has also encouraged connectivity and technology sharing for wearables and smart devices, wearable apps, and the supporting systems, aiming to bring the industrialization of wearables into reality.

Local provincial and municipal governments responded to the state policy with detailed implementation guidelines to encourage the growth of the wearable industry and the wearable market, boosting the economy by using wearables as a new growth point and creating new strengths in their industrial competitions. The Shenzhen municipal government released a key document establishing support for wearables. Under the "Shenzhen Industrial Development Planning for Robotics, Wearables and Smart Devices (2014–2020)," Shenzhen has laid down 39 detailed development policies, including targets and key focus areas.

The planning provides organizational leadership and financial support to encourage innovation, establish development programs, increase support and protections, foster high-skilled talent, broaden financial channels, and improve the environment for industrial development.

To provide financial support, the municipal government allocated RMB 500 million (USD 71 million) in special project funding to support the industry. It uses a variety of financial supports, including free credit, early-stage and continuing funding. Many financial tools are available, including direct funds, low-interest loans, equity investments, risk compensation, and more.

The government also committed to strengthening indigenous innovation by improving the ability to generate ideas. It supported the building of engineering labs, centers, and public technology service platforms as vehicles for innovation efforts. Direct subsidies were offered as special funding to the related investments. Qualified city-level engineering labs, centers, research centers, and corporate technology support centers could receive a maximum of RMB 5 million (USD 714 thousand) in special funding. Businesses, universities, and research institutions that support the building of national engineering labs or national project labs which were located in Shenzhen can receive a maximum of RMB 15 million (USD 2 million) in support funding. The building of an open or specialized sharing platform could also be eligible for up to RMB 5 million in special project funding.

In terms of attracting high-skilled talent and teams, a "peacock scheme" of special funding is deployed to focus on areas of robotics, wearables, and smart devices. Those who are qualified as high-skilled professionals in these industries can enjoy favored policies in housing, spousal employment, schooling, and academic research subsidies. These qualified talents will also be compensated according to their contribution with special bonuses and awards.

Under these policies, Shenzhen is becoming an international industrial base for robotics, wearables, and smart devices.

PART 2

CASE STUDIES

CHAPTER 7

XIAOMI – A NEW WEARABLES MODEL

On July 22, 2014, Xiaomi launched its second product, the "Mi Band". The Mi Band featured great performance and good value for money, selling at a low price of RMB 79 (around US$ 11). The Mi Band presented users with an estimate of calories burned during exercise and soon became a favorite among the young.

7.1 Xiaomi's Fan-Based Marketing

Fan-based marketing has been a key element in Xiaomi's success. Scaling production with a pre-sale stage is what Xiaomi uses to promote products. The pre-sale stage typically involves press releases, social media marketing, and offline marketing. The press release is one of the most important steps in Xiaomi's marketing efforts. During a high-profile media release event, Lei Jun, Chairman of Xiaomi, would typically give a presentation on the new product before an audience of component suppliers, fans, media, and public opinion leaders. The purpose would be to get the message out so that new products like the Mi Band would "trend" on social media.

The press release would soon be followed by social marketing events. Xiaomi always chooses the most effective and popular platforms. When Sina Weibo was the most popular social platform in China, Xiaomi carried out many "lucky draw" events involving a huge number of participants. When WeChat took over as the top social media platform, Xiaomi switched to WeChat for new product releases. The only exception was the release of the Redmi phone, when Xiaomi chose QQ space as the promotion platform, as that had greater coverage in the less-developed inland cities that were home to the target consumer base for this product. While the product supply is abundant, Xiaomi has used limited-time offers, "F-codes" (priority codes awarded to fans), and other marketing methods to deliberately create the impression of "scarcity" among consumers. Users would often voluntarily share and promote information about the product on the internet. This has proven to be a very clever and effective way of marketing.

Xiaomi's marketing efforts do not stop after achieving high volume shipments, but shift to focus on connecting users of their products to form a community. This is again very different from traditional manufacturers. Xiaomi realizes that it is possible to have a successful business when standalone device sales are less profitable or even losing money. Selling the devices cheap creates the possibility of linking products and people to provide subsequent services and derivative products which offer more potential for long-term profit. Xiaomi attempts to build such an ecosystem upon which business models are established in order to provide further services. Power companies generate most of their income not from the sales of electricity meters, but on the service of providing electricity.

How does Xiaomi operate to connect products and users? The connection is managed through the MIUI user interface system, software that forms the foundation of Xiaomi's business model. MIUI allows Xiaomi fans to connect and communicate with each other, forming a sizable community of fans. Anything generated from this community could be a potential revenue source and business stream for Xiaomi. Investors also recognized the commercial value of this community when they evaluated Xiaomi's commercial worth. And this community continues to grow every day.

7.2 Mi Band's Five Key Success Factors

How did Xiaomi convince users that the Mi Band was better than other activity trackers? Xiaomi offers some compelling answers.

1. **Unique functions.** On top of what you would normally expect to find on an activity tracker, Xiaomi's user ID recognition function has improved user retention. You could unlock a Xiaomi smartphone with your Mi Band, for example, a safe and handy function. In fact, Xiaomi was not the first to come up with this idea. Both Samsung and Google have thought of it and applied the tech to their smartwatches. But due to a lack of promotion and incompatibility issues, most users have assumed that this function originated from Xiaomi.

2. **Personalization.** Xiaomi's trackers can be personalized with an engraving option. This offering is not to everyone's taste, as many people (myself included) would not choose to have anything engraved on a wearable device. However, the market shows diversified needs around this. Many companies like this option when making bulk purchases to boost corporate identity or their company brand. Indeed, some retailers have benefited from this option by charging a considerable premium for it. It is worth noting that engraving is an option on Xiaomi's trackers only because they are one of the very few manufacturers using metal cases.

3. **Social media connectivity.** Xiaomi's Mi Band has a WeChat interface through its app. Many families buy Xiaomi's trackers together. WeChat offers a ranking system between Xiaomi users who are "friends." Family members have a lot of fun competing with each other, in terms of their step count ranking—people have even been known to cheat occasionally by tying the trackers to pets and kids. This created a social buzz in China because it integrated with the ubiquitous WeChat.

4. **Handy incoming call alert.** This function is a favorite of those who are not always on their phones. They often switch their phone to silent/vibration mode when at work and miss important calls. Mi Band's alert function addresses this problem perfectly. Xiaomi's activity trackers did not have this function initially, as the vibration motor was originally only designed for its silent alarm clock function. The vibration is strong enough to wake a person from sleep. Users found this function very useful and soon asked Xiaomi to apply the alert to phone calls.

5. **Mi Band is great for seniors.** Xiaomi's trackers perfectly match the desires of seniors in China, combining the aspiration for healthy living with activity tracking functions. We see retired couples using the trackers together on a daily basis, keeping an eye on each other's step count and sleeping patterns. These users have displayed a stronger stickiness than younger users. Older users also tend to be more into sharing ideas among friends, particularly on topics related to health and fitness. Their enthusiasm definitely contributes to the popularity of Xiaomi's products in China.

Smart devices like Mi Band have become a success story not only because the market demand is strong, but also because its well-considered ideas have proven a winning bet.

7.3   New Trends in Wearables Led by Xiaomi

Mi Band 2 was the most popular product sold on JD.com during the June 18 shopping festival in 2016. It ranked number one in product searches, beating out devices from Huawei, Apple, and Meizu. Although the retail price of Mi Band 2 increased from RMB 149 (USD 21) to RMB 200+ (USD 29+), it still sold out quickly. There were over 1.6 million pre-orders for Mi Band 2, an incredible phenomenon.

The extreme success of Mi Band 2 in China outshone the rest of the global wearable market. Jawbone, one of the earliest manufacturers to enter the market, had already ceased its activity tracker business. Morgan Stanley also lowered their forecast for sales of the Apple Watch 2. The future of wearables may be uncertain, but Mi Bands are just unstoppable. After breaking the 20 million shipment mark, Xiaomi became the second largest wearable manufacturer on the planet, beating out Apple and many others. Mi Band 2 inherited the same winning traits as earlier Xiaomi products and remains extremely popular.

Xiaomi's success story makes us want to know what product features appeal to users. The improvements in functions and the marketing success of Mi Band 2 may reveal the future of the wearables sector. We can see the following future trends:

Trend 1: Activity trackers and smartwatches are merging together.
One of the major differences Mi Band 2 unveiled is the use of an OLED display. As it presents a watch face, it could be classified as a timepiece. Xiaomi is not the first to add a screen to its activity tracker, as Huawei's tracker Honor Zero was also equipped with a screen in the shape of a watch dial. Jawbone, however, decided to neglect customer feedback and not add a display screen to its trackers, which could be one of the main reasons for its waning popularity.

As with anything, there are pros and cons. Having simpler sensor functions and traditionally no display screen, trackers tended to have longer battery life and cheaper selling prices. Considering the technology is advancing and costs are lowering, we believe more functions currently existing in smartwatches will be adopted by activity trackers as an overall trend, offering a combined solution with the best functions of each.

Trend 2: Devices are becoming stand-alone, with no smartphone needed.
Mi Band 2's revolutionary display enabled an individual to view their sensor data directly from the device, while earlier models, like most other trackers on the market, required a smartphone app for complete functionality. Since the very beginning of wearable devices, smartphones have been indispensable. A wearable device has been seen as more of a smartphone accessory with significant limitations.

When calories burned, steps, sleeping patterns, and heart rates are readily accessible from the device itself, wearables should soon be able to break away from smartphones. Instead, more wearables joined the mobile communication family—including the Apple Watch 2—in having a phone function. The TicWatch 2 and 360 Kids Watch both had independent phone functions as well.

Trend 3: The production chain is seeing greater optimization.
The wearables market is still in its infancy, with the upstream and downstream elements not yet fully developed. Parts suppliers are scarce and have low production capacity, causing a shortage of end products. The shortage of Mi Band 2 proves this.

CHAPTER 8

360 KIDS WATCH

The 360 Kids Watch is a product purposefully designed for children and their parents. When the device is worn on the wrist of a child, it is simultaneously linked to a smartphone app, allowing parents to easily locate the child. This product positioning has driven improvements in functionality through its various models over time.

8.1   New Functions Create New Demands

Precise location tracking is made possible by embedded GPS, WiFi, base station, and gravity sensors. Whether one is indoors or out and about, the watch has no problem identifying its precise location.

The 360 Kids Watch has a Chaperone Mode, which uses the smartphone's Bluetooth 4.0 compatibility to monitor the watch at a closer distance. If a child wearing the device goes beyond a certain distance from his or her parent, the watch sends an automatic alert to the mobile app, making parents immediately aware.

Realtime Voice Recording is another useful function. This the child to record short voice recordings of about 10 seconds to communicate with the linked smartphone.

In addition to these functions, the standard pedometer function is built in to record a child's daily activity, encouraging a more active lifestyle.

There are also e-stickers, a reward function that allows parents to send encouragements to the child electronically using the smartphone app. The child can gather rewards to display on the watch face.

When there is an emergency, a child can use the SOS button on the watch to send a distress message to their parents immediately.

The watch is made with non-toxic, child-safe materials such as organic silica gel and other food grade materials.

8.2   Evolution of the Functionality of the 360 Kids Watch

It has been over four years since Zhou Hongyi started working on the first 360 Kids Watch. The first three generations of the device were unremarkable. It was not until the fourth generation that it started to approach the vision of what Zhou truly wanted to achieve: a specialized wearable for children. He realized that focusing on parents, and the functionality the device offered in terms of control, was no longer what the market was after. The core function of the product had to be communication.

The device needed to evolve: it needed to become more like a smartphone.

Zhou himself has admitted that the first two generations of the product were not that successful, because 360, as a software and internet company, had a hard time learning how to make a physical device. They thought it would be easy. They learned their lesson and paid their dues as an internet company tapping into the wearable industry. Zhou made multiple changes in the development team to accomplish his vision for the 360 Kids Watch. From a quick overview of how the 360 Kids Watch changed, we can see the evolution of the product from a simple device to one that leads in its market. Examining the design of the products, the first generation focused on keeping the child in range. The second generation was more like a " detectophone" to spy on the child. With the coming of the third generation, the 360 Kids Watch repositioned itself towards the smartphone end of the spectrum—and innovations including color screens made them more attractive when the 3S model was released.

Children's tech product sales started to pick up rapidly in 2015 when the "phone watch" idea became mainstream. Neither of the previous two generations of 360 products had a phone call function. It is ironic that the developers worked hard to distinguish their product from any existing smartphone, yet failed to find success. But when they eventually added that function back, it found a warm reception immediately. What happened to smartphones was also replicated with smartwatches. Putting a color screen on the 360 Kids Watch 3S made it extremely popular in early 2016.

1. Software concept: From safety to a sense of security
"Whatever we do, we do it from the perspective of safety."

This is what a 360 employee involved in the promotion of the Kids Watch products said in early 2014.

The market response to the first two generations of Kids Watch products did not please 360. The focus on child safety was not well-received and the product was not selling. Safety was what Zhou Hongyi had in mind when he came up with the concept, but this placed a bit of a straitjacket on product developers. Zhou eventually had to change the definition of "safety." During the new product release, he explained that "this watch is not a product to solve the problem of lost children, it also addresses the issue of feeling safe in a general sense. This means the ability to communicate with children and be connected to them."

In general, parents want their children to be safe, but safety is only a baseline. It cannot be seen as a core value. From a product management point of view, the target consumers of the first two generations of products were the controlling parents, not the kids themselves. Later generations of the product focused much more on the children, with parents' concerns considered. By now, knowing what the children really wanted proved to be the key to success. In a practical way, in weighing the genius of a product developer with the precise analysis of user behavior data, a company should always give more credence to the latter in determining the direction of product development.

Zhou Hongyi still sticks to his internet business model and online channels. His products remain at a relatively low price point, regardless of competitors' high price tags and their focus on developing offline channels. He eventually saw the fruit of his efforts.

8.3 Smartwatches for Kids Take the Lead in China's Wearables Market

Unlike the rest of the wearables market, smartwatches for kids are thriving, with rapid growth in China. The children's market is a battlefield for all players in wearables. Early in 2016, we saw many companies entering the children's wearables market, including BBK, Baidu, Huawei Honor, Tencent, Li Ning and Sougou. Wang Liyang, an internet analyst, said in an interview with National Business Daily, "Safety and companionship are the most noted pain points of the children's wearables market. The market demand is very clear and young parents are willing consumers with special needs in childcare. This is why major players are competing relentlessly in this market segment."

1. Starting from content
Research shows the global wearables market will exceed RMB 10 billion (USD 1.4 billion) in 2015, and likely reach RMB 2.5 trillion (USD 357 billion) by 2017.

The boom in wearables is partly a result of the recent AI revolution. However, as with most devices, wearables are limited by the content supply due to operability issues. LeTV has refocused itself to build on its strength in content, after a six month exploration of the market.

The newly released Kido Watch reportedly collaborated with content providers such as ximalaya.com and 61.com, which provided personalized nursery rhymes, bedtime stories, science readings, Chinese culture, English language and many other educational and entertainment content. The Kido

Watch also connects with tuling123.com, which provides a search engine to reply to children's questions in a child-friendly manner. Children can use the device's voice recognition feature to ask questions about nature, math, the weather, and more.

2. *Industrial giants wrestling for their market positions*

There is no doubt that the children's wearables market shows strong current demand and also has tremendous potential for future development.

In 2014, Sogou.com released its Teemo smartwatch for kids. Meanwhile, the Okii smartwatch for kids developed by the BBK Group achieved 700K shipments in 2016, ranking it fifth among wearables in total global shipments that year.

According to IDC's quarterly report on the global wearables market, from July to September 2015 Okii saw more shipments within China than Samsung's wearables saw around the globe.

After industrial giants, entrepreneurs, and capital investors all rushed into the market, more variety has appeared in children's wearables, including smartwatches, smart shoes, and even smart clothing. In 2016, Tencent announced its collaboration with Lining, Daphne, ABC KIDS, and other shoemakers to launch QQFind, a line of smart shoes for children.

And yet, people still have doubts about children's wearables. Wang Liyang thinks the products currently on the market are quite homogeneous. It is difficult to see ground-breaking innovations when most players are traditional industries collaborating with internet companies for an upgraded business model.

Wang also observes that this uniqueness limits product sophistication. Teemo T2 and Huawei Honor Small K are both excellent at indoor location detection, with an accuracy within 20 meters. Outdoors it is accurate within 5 meters, which is also remarkable. To prevent a child from going missing, this is useful. But in an abduction scenario, once the device is removed it is immediately of no use. The extension of story and educational content is still limited by the physical size of the screen. This is another area unlikely to see major breakthroughs in the short term.

CHAPTER 9

FITBIT TRACKERS

Fitbit held the lead position in the global wearables market in the first quarter of 2015, with 34.2% of the entire market and total shipments reaching 3.9 million units in that quarter—a year-on-year increase of 130%.

In May 2015, Fitbit, the activity tracker manufacturer, submitted its IPO documents to the US Securities and Exchange Commission, aiming to raise a total of one hundred million dollars.

In its IPO prospectus, Fitbit reported revenue in the first three months of 2015 (ending March 31) up to USD 337 million, and a net profit of USD 48 million. In 2014, Fitbit sold 10.9 million trackers for a revenue of USD 745 million and net profit of USD 132 million. Revenue doubled that of 2013, and 2014 marked its first profitable year. In 2013, the total revenue for Fitbit was USD 271 million, with the net loss of USD 52 million. Further back in 2012, the total annual revenue was only USD 76 million, with a net loss of USD 4 million.

On June 18, 2015, Fitbit's stock began trading on the New York Stock Exchange. It opened at USD 30.4, surging nearly 52% from its IPO price of

USD 20 per share. Fitbit sold 36,575,000 shares in this IPO, and was already worth more than USD 6 billion on its first trading day.

So what has Fitbit done right so far? Let us take a look at its company history, up until 2015.

- October 2010: Fitbit established
- October 2011: Debut of the Fitbit Ultra
- January 2012: Round C financing received, valued at USD 12 million
- April 2012: Smart scale Fitbit Aria released
- September 2012: Fitbit One released
- September 2012: Fitbit Zip released
- May 2013: Fitbit Flex released
- August 2013: Round D financing received, valued at USD 43 million
- October 2013: Fitbit Force released
- Early 2014: Fitbit Force reported to contain nickel, causing skin irritation among many users
- October 2014: Debut of Fitbit Charge, Charge HR, and Surge—their first phone-connected device with call and text alerts
- November 2014: Apple removed all Fitbit products from its retail stores
- May 2015: IPO application officially submitted

The wearable device company started by manufacturing belt-clipped pedometers. They launched over 10 models of wearables over 8 years, raising USD 66 million in total.

Fitbit now has more than 4,500 physical outlets in over 50 countries and has an online channel for worldwide sales, reaching every corner of the planet.

9.1 Fitbit's Five Key Success Factors

We have seen that many companies and products have come and gone in the wearable industry in a short period of time. Only a handful of players have been resilient and thrived. What makes Fitbit one of those very few winners? Let us look at the major factors of its success.

1. Profitable products are the only good products
Fitbit generates its revenue in two ways.

First, the selling of devices. With limited technological resources, Fitbit focused on targeted market segments to develop suitable products. Profits generated from device sales were then be used to fund further R&D for better products and services.

Second, Big Data has helped to bring in profits. Reliable products have enabled the accurate tracking and monitoring of user data. The mining of high-volume, accurate user data allows for further analytics and studies by individuals and institutes. This becomes an additional profit-generating business stream.

2. Actively informing consumers "who I am"
Some sellers of wearable technology struggle to clearly explain their products to consumers. Popular products like smart glasses, smartwatches and activity trackers are all well-marketed. Adding the term "smart" in front of traditional items, in and of itself, has given the straightforward impression that this is a more technologically advanced version of the product. Consumer education is a difficult area. Tech start-ups struggle when they try too hard to convey more complicated concepts.

Fitbit always uses easy words to position its products, simply telling consumers "who I am." "I am a smart wristband," "I am an activity tracker." This is also marketing. Fitbit puts everything in a nutshell to let consumers know that their products make you fitter through the appropriate support of technology.

3. Fewer functions go further
Many wearables today are overloaded with functions, making it harder for ordinary people to use them. These ever-more-complex gadgets are more suitable for geeks than average consumers. Chinese industrial development at

this stage is not mature enough to support all-in-one products. Users are not ready to accept this concept either.

As one of the earliest entrants into the market, Fitbit managed to commercialize its business and focus its products solely on the fitness segment. It has developed vertically along this market segment to great success. Combining exercise tracking with heart rate and sleep monitoring, it didn't try to fit many other functions into its devices. Instead, Fitbit focused on simplification, with minimum functionality. Product issues have therefore been easier to manage and resolve, and performance has been improved for consumers.

4. Focusing on one market segment

Focus is not achieved easily, as this sometimes goes against human nature. We always want to have more. But Fitbit has consciously used focused products to effectively attract a targeted user group. Throughout its history, Fitbit has focused solely focused on wristband-based activity tracking and fitness monitoring devices. Based on this strategy, Fitbit regularly releases updated versions to meet consumer demands.

5. Data value generation

Most decision-makers in the wearable device industry contemplate the commercial value of the Big Data collected by their products when considering their business models.

To be honest, there is still very little hope for such commercial value to be realized in the next few years in China. Most wearables made overseas would last a few months at most before being left at the bottom of the drawer, while our Chinese products are not used more than a few weeks on average. Data collection is very fragmented, therefore most of it is of no value at all.

Insufficient data collection leads to a dilemma of what to work on first. Sufficient data could support the development and refinement of algorithms to facilitate more precise tracking and more accurate feedback for users, and in so doing lead to increased user retention. On the contrary, the inadequate data collection leads to inaccurate monitoring, which will inevitably result in user dissatisfaction—and as a consequence, algorithms become more hypothetical, and user stickiness diminishes. This is a bit of a chicken and egg problem. In the next section we will look at the risks for Fitbit.

9.2 Risks Faced by Fitbit

Having identified the major factors behind the success of Fitbit, it is important to examine the risks faced by the company.

In November 2014, when Fitbit was a thriving player in the wearables market, Apple suddenly decided to remove all Fitbit products from its retail outlets. Apple did not offer much explanation for this action, but most believe that was for two reasons. One reason is that Fitbit did not support Apple's HealthKit platform—and this still holds today. The other reason is thought to be the remarkable similarity between the then-newly released Fitbit Surge and the Apple Watch, including a touch screen, eight sensors, in-coming call alerts, text message alerts and more.

Therefore, it is said the major risk for Fitbit is coming from Apple, as Fitbit refuses to join the Apple fitness platform. Some suggest Apple did not like this so it kicked Fitbit out. At the same time the Apple Watch emerged as the ultimate replacement of early-model activity trackers. But one could take a different view of the situation.

According to the statistics Slice Intelligence provided to Bloomberg, Fitbit sales in the first quarter of 2015 were not affected by the new Apple products. Apart from the first week following the Apple Watch release, Fitbit outsold it in actual shipments most weeks.

Comparing the overall strategic positioning of Fitbit with the Apple Watch, it is reasonable to suggest that Fitbit chose the right one from the very beginning.

The underlying cause of the separation from Apple was a conflict over the platform. Fitbit could not remain successful without its own platform, which could be open to all developers.

Why? First, Fitbit needs the attention from users and device makers to enable the development of new products and creation of more interesting user experiences. Next, Fitbit benefits from the ideas and innovations generated by these developers. The provision of better products, assisted by more reliable, more personalized and unique services is crucial for user retention. Concerns over the control of data is merely a passing issue.

On the other hand, Fitbit cannot be seen as a direct competitor of Apple. Apple is wrestling with the traditional wristwatch industry in Switzerland, and this is the true market battlefield where the giants are competing. Fitbit focuses on a small and specific market segment, which is not quite the same challenge.

In terms of functionality, in the short term there will not be much overlap between the users of Fitbit and the Apple Watch. Again, this is because the focus of Fitbit is on sports and fitness, and its related vertical segments, while the Apple Watch has a much wider functionality. According to Slice Intelligence, only 5% of people bought the Apple Watch after buying a Fitbit in 2013 and few have moved the other way, with about 11% of people buying a Fitbit after using the Apple Watch.

The biggest risk for Fitbit is not the actions of Apple or other similar producers, but from within Fitbit itself.

What made Fitbit could also break it. And the reason is simple: the exercise and fitness market does not exhibit the condition of rigid demand.

If you look at the two types of potential users, the first group includes those who love fitness and exercise, while the other contains those who are not fitness fanatics. For the first group, fitness trackers do not have much meaning for them. Such people exercise anyway, with or without a tracker. The daily targets set by Fitbit, such as number of steps or calories burned, are easily attained. As such, it is difficult for this group to develop a reliance on their Fitbit devices. The second group of people, on the other hand, may like Fitbit initially, with its gentle reminders to get up and move. But they are also very easily disappointed when they find the additional exercise doesn't quickly achieve their targets, leading them to soon lose patience with the device.

"42% of people stop wearing fitness trackers after six months," Wall Street Journal technology columnist Christopher Mims wrote on Twitter. "Any growth left for Fitbit is just people who haven't tried or discarded it yet." Both groups of users share one similarity, though. They are both easily attracted to devices with newer functions, such as smartwatches.

9.3 Two Options Ahead for Fitbit

When analyzing Form S-1 for Fitbit's IPO, Redpoint Ventures pointed out that there are many activity tracker makers in the market but very few are truly of scale. Jawbone was the second largest activity tracker manufacturer at the time, but they were troubled by internal restructuring and staff retention issues. Many of their employees left to join Fitbit. Nike had the potential to be the third largest player, but in 2014 they decided to cut their 70-person hardware team and focus solely on software. The rest of the products are all from small teams, many of which use Kickstarter and Indiegogo for crowd funding. Fitbit is undoubtedly the dominant player in terms of both volume and market share.

As successful as Fitbit has been in the activity tracker market, the IPO still raised many questions. Bloomberg even warned that Fitbit should be very careful not to become the next Blackberry.

I believe the future of Fitbit is in their hands. Nokia did not collapse overnight, and Fitbit is not necessarily going to follow in their footsteps. I therefore would like to give two suggestions for Fitbit's future positioning after the IPO.

One is to go deep in the vertical market, for example to act as a fitness instructor or coach for users that want to pursue yoga or other activities. Indeed, we can tell that Fitbit has been moving in this direction. In 2012, it launched its smart scale Aria. In 2013 it added Fitbit Force to the family with a small OLED screen to its activity tracker band, though it was later recalled due to its material. In March 2015, just a couple of months before its IPO, Fitbit acquired Fitstar, a personalized fitness coaching app, at the cost of USD 17.8 million. This app offers personalized fitness coaching lessons, which brought many new users to Fitbit Coach.

The other suggestion I would give is to move to the medical wearables market where rigid demand exists. Fitbit could use the collected data to analyze user health conditions. For those with chronic disease, it could offer a special tracking service. Through data processing these users could be offered personalized health wearables to enable them to better manage their diseases and lifestyles.

If Fitbit focuses only on fitness instead of expanding to the above two areas, I dare say that it could potentially face fatal problems sooner or later. After all, fitness trackers are not sophisticated enough devices to create a technological barrier. If industrial giants like Apple decide to move into this field and take over, Fitbit will not be able to retain its leading market position for long.

CHAPTER 10

NIKE+ JUST DATA IT! NIKE'S WITHDRAWAL FROM HARDWARE TO SOFTWARE

In 2012, Nike made a strategic decision to expand its business from sports equipment and clothes towards the fitness technology market. It launched its first fitness tracker, the FuelBand, in the same year. They created the brand "Nike+" to bring all their fitness apps and devices together. Apps under this brand include Nike+ Running, Nike+ iPod, Nike+ Move, Nike+ Training and Nike+ Basketball. Wearables include the Nike+ SportWatch, the Nike+ FuelBand, the Nike+ SportBand and others.

Nike+ Sneakers can be wirelessly connected to an iPod through Apple's Nike+ iPod suite, and the iPod can be used to store and display exercise data including date, time, distance, calories as well as total exercise reps, time, distance, and calories burned. Such real-time data could also be conveyed through earphones or computer speakers with iTunes installed.

FuelBand was officially released in February 2012, and was warmly received. In November 2013, Nike released an upgraded SE version of the device. Nike announced that the total number of users on its Nike+ platform exceeded 18 million in August 2013. This figure increased to 28 million by April 2014.

Meanwhile, Nike started to explore other training products, such as the N+ TC for female fitness training and the Nike+ Move.

However, the story had a turning point in April 2014 when Nike dramatically decided to withdraw from the wearable tech market. They explained that after much consideration, they believed that the fitness software business had much better prospects. Therefore, all development on the FuelBand and other wearables was abandoned, and over seventy employees made redundant.

Fitbit and Nike were both early movers in the area of fitness and sports wearables. How has Fitbit, a 2007 startup, come to dominate the market, leaving the more established Nike in its wake?

10.1   Reasons for Nike's Exit

Nike's dramatic exit can be explained by four critical developments, as described below.

1. Apple, Nike's ally, entered the wearables industry

Nike was one of the pioneers in the wearable technology field and had a very good overall performance with its fitness tracker products in terms of sales, profitability, and market influence. When it announced its decision to quit that line of business, many believed that it signaled the wearable tech industry was not looking so promising any more. And yet the true reason behind this move was because of Apple's entrance into the wearable tech market.

Looking at the shareholders of the two companies reveals a friendly and deep relationship between the two. Previous Nike products also benefited from Apple's global sales channel. When Apple announced its strategic move into wearables with the Apple Watch, Nike chose to exit this market and turn instead to research and development on the use of data from wearables and the creation of algorithms to explain behavior. The tacit understanding between Nike and Apple has led Nike to refocus to capitalize in a different way on its rich experience in the sports and fitness industry.

2. Nike over-estimated the production chain behind wearables
As a corporate giant and leader in the sports industry, Nike is more than capable of developing a specific technology to complement its products. For instance, Nike single-handedly developed Nike+ Force Sensor technology to be used in its sneakers. This technology allows the wireless transmission of exercise data from the footwear to other mobile devices, as well as sharing through social media. What Nike overlooked was the chain of other related technology required for the marketing of wearable devices. Apart from sensors, there were batteries, data uploads, information exchange, designs, Big Data analysis, and many other indispensable areas. There wasn't yet a complete ecosystem to form a working production chain to feed all these areas. Nike could not resolve all these aspects on its own.

Nike's strength has always been in sportswear, not in electronics. It always seemed slightly out of its depth when tackling the technical issues involved in the market for smart electronic devices. Apple, on the other hand, sits on top of the world of smart devices. Apple is able to leverage its technical and brand influence in the smart industry as well as its supply chain and management abilities to acquire the highest quality technology available.

The smartphone industry has a much more mature industrial ecosystem at present compared to that for wearables. It may be a long time before the wearables industry catches up. If Nike were to choose to collaborate with Apple, possibilities would open up to produce new smart products to work with Apple devices, such as iOS-supporting smart shoes or smart sports clothes.

3. Nike overestimated the returns on the business
In an analysis of 2013 sales data, Fitbit, Jawbone UP and Nike FuelBand accounted for 97% of all activity trackers sold from major online vendors. From early January 2013 to January 2014, Fitbit devices sales accounted for 68% of the total, while Jawbone UP only had 19%, and Nike FuelBand merely 10%. Nike was already lagging behind.

There were intensive debates within Nike's management team. As the FuelBand hardware business was expensive to run, and the production chain was facing difficulties, the business could not bring in enough profit. This led to debate over the prospects for this business line and the need for decisions. There is some suggestion that, during the development of FuelBand, Nike was never quite able to attract the high-level engineering talent that was needed. Nike's dilemma was a situation faced by many. Most companies are willing to try out new ideas but are reluctant to invest enough resources to go beyond the trial stage.

Nike was also restricted by its sales channels. They very much relied on Apple's channels at the beginning, but when Apple itself decided to enter this field, Nike wearables lost out. With costs still rising and market share dropping rapidly, grim prospects forced Nike to make the decision to close down its wearable hardware business line.

4. More potential in software

"The agenda isn't to sell FuelBands," said Jim Duffy, a Nike analyst with Stifel, Nicolaus & Company, on Nike's wearable focus. "It's to develop this customer database and pool of customers that they can have a dialogue with… to make people more active and thus increase demand for core products."[2]

What we can see from this is first that Nike does not plan to entirely quit the wearable market. On the contrary, it envisages greater business potential in areas other than hardware. It is obvious that data management will become a crucial part of the wearable market as device makers all need to collaborate with third parties in that area. Second, the existing hardware market for activity trackers was already quite crowded with Fitbit, Jawbone, Withings and Garmin. The competition was incredibly fierce. Nike was not the strongest in hardware, plus the technology, supply chain and marketing of digital devices are entirely different from their core business of sneakers. It proved too expensive to keep it going and with little chance for success. Third, the Nike+ digital fitness platform is well developed, setting Nike up to be a data collecting company and an app maker. Instead of selling hardware, the massive sports data set Nike has may bring the potential for significant revenues.

After a deliberate restructuring of the hardware team, Nike made its intentions clear. It turned away from hardware development to focus on its resourceful core sports business. It would go on to create a business ecosystem based on Nike+ and to provide digital services to other wearable makers.

10.2 Nike's Software Strategy

In March 2015, just before the launch of the Apple Watch, Apple removed Jawbone UP and Nike+ FuelBand from its retail stores. Simultaneously, Apple made friendly overtures to the sports company by adding a Nike+ Watch app to its list of recommended fitness apps. This app, Nike+ Running, allows Apple Watch users to connect with Nike's global running community while recording distance and time on Apple's device.

Nike+ Running is a good app for runners. It is easy to use with the Apple Watch, with simple functions like maps, music (added via smartphone), recording running distance, and providing the time. This data can be synchronized and managed together with other health data collected on the iPhone. By no means a complicated app, Nike+ Running allows users to leave their phone behind and still view their running data.

In fact, the collaboration between Nike and Apple on wearables has always been close, going back to 2006 when the two jointly developed Nike+ iPod. It was a very popular fitness tracker. Apple's CEO Tim Cook is currently a member of Nike's Board of Directors, and has been for a number of years. Apple even hired the former Design Director of Nike to be in charge of R&D and design for the Apple Watch. Before that Nike also had a long-term relationship with Steve Jobs. The common understanding of the wearable tech market today is that fitness products are the easiest route to get into the field. Nike is the leader in sports equipment, and Apple is a dominant force in the smartphone business. Partnership between these two allows them to leverage the strength of both brands and their respective power in technology and service. It opens up business potential that no other pairing could reach. Mark Parker, CEO of Nike at the time, revealed that part of the collaboration between the two companies is to explore the digital frontier of wearable tech, and to develop the Nike+ user base.

Although its wearable development unit has closed, Nike continues to work on fitness apps and the upgrading of existing products. Parker revealed that there are over 60 million users of Nike fitness apps. He also said that wearables are important to Nike, and they are the center of the Nike brand.

It is unclear how Nike and Apple will continue their partnership in the future. They may launch new products together, or they may only work together on the Apple Watch. Time suggests that three functions are essential in any collaboration between the two:

1. Smart music management

Nike's iPod and iPhone apps are both available for playing and controlling music. Considering that the Apple Watch is also able to control music played on the iPhone, Nike could add a smart music control function with its own sports features. They could group music by tempo, to play music in same tempo as the exercising pace. Slower tempos would be useful for jogging, and faster beats for more intensive movement. If such a function could be combined with the play history and preferences of the user, the result would be even better.

2. Supporting more activity types

This is an obvious direction as previous Nike apps were only focused on running. When the wearables are connected to a mobile network and GPS, more activity type support is possible. Cycling and golfing are two good potential additions. Exercise performance and distance measurements could be used with cycling, and weather monitoring could be added with golfing.

3. Deeper integration via social media

Some social functions have already been incorporated into Nike's apps, which enable competition among friends. With its new functions, the Apple Watch is further along in this area.

The Apple Watch had a new response function called "taptic" which created a gentle tapping on the wrist. This function could be used as starting guns to send signals to people at different locations simultaneously for a race, or to give feedback when the user is overtaking competitors or lagging behind.

It is foreseeable that Nike's strength as a sports brand could be vertically integrated with wearable apps, just as Google does in the medical industry. This is where the wearable tech industry is heading—more segmentation and a vertically integrated ecosystem.

Overall Nike and Apple have chosen the "division of labor" model for their strategic collaboration. Hardware is given to the electronics industrial leader Apple, and sports data is handled by sports expert Nike. Nike is therefore able to maximize the value of its accumulated knowledge in the sports area. Being more specialized and deep-rooted in such an area makes it easier to create a technical barrier against others.

Apple also needs Nike's help for better performance of wearables' sensor functions. To maximize commercial value, it was a wise move for Nike to forgo the hardware side of the market and turn to focus on software. Nike will also drive forward the entire wearable tech industry in developing sports-based sensor technology.

CHAPTER 11

GOOGLE GLASS

In 2012, Google shook the entire tech world, and even the world of fashion, with its revolutionary Google Glass. Sergey Brin, co-founder of Google, wore the device as part of New York Fashion Week. But in 2013, Google Glass was banned in many public places in Seattle and the U.K. because of privacy concerns. the problem further escalated in 2014 when a person wearing Google Glass was attacked in San Francisco. Negative reports about Google Glass became rampant. In January 2015, sales of the Explorer version of Google Glass were halted, leaving many in the tech world disappointed. The three-year roller coaster ride following the birth of Google Glass had many heart pounding moments.

In the three years since its unveiling, Google Glass upgraded its functionality to include all types of lifestyle needs, including notices and alerts, weather, voice control, traffic monitoring, maps, GPS navigation, point of interest details, a camera, video chat, gaming, real-time translation, and more. From the range of functions, we can sense the broad ambition behind Google Glass. It has applications across multiple industries including transportation, mapping, tourism, advertising, and streaming media. What does Google Glass really want to become? What does Google want to achieve through Google Glass?

11.1　Google Glass in the Consumer Market

On April 4, 2012, Google announced its digital glass program called "Project Glass" on its social network platform Google+. Following this, on February 20, 2013, Google offered a competition for those keen to test the Google Glass. The application deadline was February 27 and only US citizens over 18 years of age could apply, for a fee of USD 1,500.

On October 30, 2013, images of the second-generation Google Glass were released on Google+. The first generation had adopted Bone Conduction Transducer (BCT) technology to transmit sound. The second generation added earphones.

On April 10, 2014, Google announced the online sale of Google Glass scheduled for April 15, with any US residents over 18 eligible to purchase one at the price of USD 1,500. The sales window was only one day.

On May 25, 2014, Google started to sell the Explorer Edition of Google Glass to all Americans over 18 years of age. It could be purchased directly from the official Google website.

On June 23, 2014, Google announced the first international launch of Google Glass, now open to residents of the United Kingdom. Its selling price in the UK would be GBP 1,000 (USD 1,713 at the time).

On November 25, 2014, Google decided to close its physical retail shop Basecamp, where Google Glass was sold. Most users were using the internet or telephone for purchases and technical support.

In January 2015, Google disclosed that Google Glass would be transferred from the Google X department to the consumer product department. They announced the closure of the Google Glass Explorer program and the halt of the product sales.

Google Glass had, it is fair to say, a few rough years. It was either challenged for not being aesthetically pleasing or boycotted for privacy issues. In fact, before the launch of the product, Google Glass had gone through many prototypes at Google Labs. It transformed from early invisible models to its eventual glasses format. But the entire Google Glass program was seen as a go-to-market experiment. On March 23, 2015, Google CEO Eric Schmidt issued a statement that Google would continue the developmental of Google Glass, as this technology was too important to give up.

As one of the keynote products of the entire wearable tech industry, the ups and downs of Google Glass have attracted tremendous attention. It popularized the concept of wearable tech and provided substantial impetus to the Internet of Things. Google Glass encapsulated a range of things, and its impact was felt beyond the product itself. Its meaning transcends itself – opening a new vista for business potential.

11.2　Cap-Mounted Google Glass Could Lead the Way

According to media reports, Google has obtained a patent for mounting Google Glass onto a cap. As illustrated in the patent, the device is made up of a cap connector and a display. The display could be attached magnetically onto the cap, moved to different locations, or rotated to different angles.

Since the debut of Google Glass as a physical application of a smart wearable device, the wearable industry has been on a bit of a roller coaster ride together with Google Glass. It was produced high expectations, then faced criticism and controversy. And yet Google Glass can stand tall as a ground-breaking product with a major role in ushering in the new era of wearable technology.

In fact, the latest Google Glass is a remarkably fine product in terms of technology, appearance, and interactive functionality. It is a cutting-edge masterpiece offering an unparalleled experience. Unfortunately, the commercial aspect of Google Glass has not been as successful, partly because the initial strategy was not focused on the marketization of the product but on the revolutionary concept itself. The failure to integrate Big Data in applications also hinders the performance of Google Glass. Many consider Google Glass to have failed, but it is possible to take a different view.

When Google Glass was first presented by Google X, it was not a prototype. Instead it was in the form of a ready-to-use product. Before it was shown to the public, it had already gone through many internal trials and tests. Imagine the concept starting from an idea, and materializing into some clumsy prototype that was barely usable, and being optimized again and again. Some of these upgrades would not be successful, and some constituted minor incremental improvements. Google kept going and supported this dream with extraordinary perseverance.

The Google Glass we have seen is only one of the many versions Google developed during this long process. The timing of the launch and the version of Google Glass chosen was based on their judgement about the technological transition to the Internet of Things, when wearables were ready to be considered as an industry. Google Glass was therefore introduced to ignite the wearable tech industry and to perform commercial-scale testing.

11.2.1   Strategic intentions of Google Glass

Initial testing started with the mass consumer market and applications in media, education, social networks, entertainment, and other fields. Then it moved on to remote medical treatment, trials in the UK market, and enterprise-level applications. Google is still looking for an optimum commercial model for Google Glass.

Previously, Google X had relied on its internal experts to perfect the product concept and refine its design. To further capitalize on its commercial value, more work needs to be done in practical application scenarios. It is important to know the limits of Google Glass and its applications, as well as specific technical requirements for different needs and uses of the device.

This latest patent acquired for Google Glass is a newer version based on user tests. The previous Google Glass versions all used a spectacles-based format, which is not preferred by some users. This new format has made smart glasses intangible, allowing their use by people not into wearing spectacles.

As an early explorer of wearables, Google has many insights in this industry. Wearables formatted as spectacles have far more application potential and far higher commercial value than today's popular smartwatches and activity trackers. In other words, smartwatches and activity trackers are in fact the product category with one of the most limited set of applications among all wearable formats. The next wave of mass marketing in wearables may be smart glasses and smart garments. Google has changed its company structure, which may help accelerate the growth of smart glasses and attract more talent to the race for developing real-world applications and a Big Data platform, driving the commercialization of the entire wearable tech industry.

11.3 Google Glass in the Corporate Market

Mao Zedong said the revolution should start from rural areas where the enemy is weak and gather resistance to surround the cities to gain final victory. This is the method employed by Google with Google Glass. The troubled route into the consumer market has made Google Glass shift its focus to the corporate market, as tech blog 9to5google.com reveals. Whether this is true or not, one thing we know for sure is that Google Glass was popular during the trial period among its corporate users. Such a strong and positive response meant a great deal to Google Glass after facing much rejection from the consumer market.

Since the Google Glass Explorer program was halted in January 2015, there have been signs that the focus is shifted to the corporate market, which could be a vital step in winning back the consumer market in the future. There are a number the reasons to support the suggestion that Google Glass may be one to keep an eye on, and in the following section we will examine them in depth.

1. Rosy prospects in the corporate sector for wearables
According to the U.S. Bureau of Labor Statistics for 2012, about 46 million Americans were employed in industries that could potentially benefit from the application of wearable tech. By 2022, this figure is expected to rise to 52 million. And this is just the United States. The rest of the world is crying out with the same demand. The following statistics further demonstrate what we can consider to be a rosy future ahead.

In fact, corporate users have always been willing to try out new devices still in their early marketing stage, including computers, smartphones, and tablets. When the wave of mobile devices first arrived, corporate users adopted this advanced technology in their operations rather quickly. At one stage it seemed like every executive was equipped with a Blackberry to help with their business. Likewise, we may be just about to see different types of wearables become mainstreamed first among those employed in the media, hospitals, schools, factories or those working hazardous jobs. Boeing has been an early adopter, for example. They introduced the use of smart glasses for their engineers so that traditional manuals could be replaced by a quick, hands-free content search through the wearables.

Companies using Google Glass and similar wearable devices could save up to USD 1 billion within three to five years. These devices may be particularly helpful in the fields of mechanics, medicine, and manufacturing where hands-free internet access, a video camera, and video streaming are most useful. Dignity Health, a US medical institution, adopted Google Glass for its doctors to use during consultations with patients. Time spent on inputting data dropped from 33% to 9%, while the time spent speaking with patients increased from 35% to 70%.

With greater business acumen, corporate users seem to be willing to try out new technology and products to improve productivity and save costs.

PricewaterhouseCoopers once conducted an influential study of 1,000 American adults. In this study, 77% of participants believed that wearable tech could open new opportunities and make people more productive in work. But just under half of participants (46%) thought their companies should invest in wearables for the staff.

J.P. Gownder, a Forrester Research analyst, also pointed out that using smartwatches and other wearables in work could improve data analysis, noting that "In the future, cognitive computers (like IBM's Watson) and voice-controlled intelligent agents (like Siri, Cortana, or Google Now) used with wearable devices will augment the skills of humans on the ground, helping them identify and act on specific problems."

Obviously, both employers and employees are keen to use wearables in their workplace. For Google Glass, the business user market is not only an ocean of promising customers, but also a high-stakes testing lab. Google Glass could either win big or lose it all. This is still a better target than the consumer market where negative opinions on Google Glass have taken root and may prove hard to eradicate. Strategically Google is wise to choose the corporate market at this moment, and so embrace the huge market potential of corporate users.

2. The business user market: segmentation of the market enables focused efforts
Mao Zedong, in his *Strategy of the War of Resistance Against Japan*, wrote that in guerrilla warfare it is important to distribute the battlefield in order to spread out the targets. The business user market is similarly a large battlefield with users of different business models, operational approaches, job types, and many other differences. In other words, the entire corporate market could be segmented into many parts or points according to different features.

Before Google Glass was launched to the consumer market, Google was also exploring the business user market. In March 2014, Augmedix, a startup that uses Google Glass as an electronic medical record solution, raised USD 3.2M in venture capital. The Dow Jones called it "the first publicly announced round of venture financing for a developer working exclusively on Google Glass," highlighting the device's potential among business users.

Google launched a "Glass at Work" program aimed at making apps for business to improve working conditions and productivity. In June 2014, Google announced the first five Glass at Work-certified partners. They were APX Labs, Augmedix, CrowdOptic, GuidiGO and Wearable Intelligence (see Box 1).

Box 1 Initial Glass at Work-certified partners

APX Labs – Creator of Skylight business software for Glass, enabling instant access to business data at work

Augmedix – Developers of the app allowing doctors to access patients' medical data, including heart rate, blood pressure and pulse rate, on Google Glass

CrowdOptic – Providers of content for live broadcasts and context-aware applications for sports, entertainment, building/security, and medical industries

GuidoGO – Partnering with museums and cultural institutions to help people connect with art and culture through story-telling and other experiences

Wearable Intelligence – Creators of Glassware for energy, manufacturing, healthcare, and more

As can be seen from the five certified partners of Glass at Work, their activities cut across many different areas and walks of life. Each of their apps have a clear purpose designed for precisely targeted professions. They require users to have certain professional knowledge. The global oilfield service giant Schlumberger has worked with Wearable Intelligence to equip its technicians with a modified version of Google Glass for quicker access to stock information and improved efficiency.

The fate of Google Glass in the business market is rather different from the consumer market. Through 2016, the Glass at Work program had attracted 10 partners developing various versions of apps for Google Glass. A Google spokesman said the company would keep investing in this program to seek more developers for business. Brian Ballard, co-founder and CEO of APX Labs, said that the sales of Google Glass to business users were taking off, noting the growth increased phenomenally every quarter. Auto companies, airplane makers, power companies, telecoms, and many other business giants signed up. APX Labs signed an agreement with Boeing in November 2014 to develop a Google Glass trial.

The success among business users cannot be taken as proof of strategic failure in the consumer market. On the contrary, it shows that Google Glass has the ability to customize itself to meet user needs. It offers clear and specific service details for business customers. Technology is not a challenge for a digital giant like Google, but knowing the customer is. Google was clearly confused about what customers really wanted before Google Glass was launched.

The business market is more focused, and Google Glass is offering point-to-point service. Success in this market could eventually lead to an overall win in the mass market.

3. How to execute outflanking tactics
Google Glass Explorer Edition had an inauspicious end, but that is a fate pioneers often face. As long as Google keeps exploring, there is always the opportunity to adjust and try again. Here we will discuss various tactics that could be employed:

(1) Stir up again and rekindle the market
As previously mentioned, the business market for wearables looks very promising. However, most wearable makers still generally focus on the consumer market. Crowd-funding websites see many gadgets developed for niche markets. These startups come and go with very few survivors. Google was there and it did not progress far either. As a consequence, Google changed direction and went for the business market.

The advantage of being a big company is that Google can always afford to be the first to try something new. With strategy and foresight, Google did not launch a blind effort. It had one positive and one negative initial result from these two different markets. Google then decided to regroup before its next move.

With the successful Glass at Work program, Google is now ready to build an industrial ecosystem of wearables. Google owns its own smart hardware device, Google Glass, a leading product among rivals. Google also has its own wearable operation system, Android Wear. Last but not least, there are increasingly more upstream and downstream players joining the game. Developers have made successful enterprise-grade Google Glass apps under the Glass at Work program. Businesses have seen the benefits of Google Glass and are paying for this service. The chain is therefore formed.

Once a thriving ecosystem is established, it will only take time before the other issues are solved.

(2) A leap of faith

One of the purposes behind the Glass at Work program was the hope that a positive impact on the business market could change the shady image of Google Glass among consumers. Google wants consumers to rekindle their faith and accept Google Glass again.

It is not an easy path to tread. The business market has more treacherous terrain than the consumer market. The PC and smartphone life cycles both prove that technological challenges and progress are likely in the corporate market.

Google Glass needs to first tackle the issue of customizing service for business clients before addressing the considerable task of information safety. Compared to consumers, businesses are more interested in device-enabled productivity improvements and procedure streamlining. The hardware configuration must be adapted to suit the frequency and intensity of usage in the work environment. It is reported that Google has made significant hardware changes to suit the needs of corporate customers. The next generation of enterprise-grade Google Glass will simplify the design in exchange for a bigger prism, a more powerful Intel Atom CPU, and a longer-running external battery.

Apart from generic business needs, Google Glass must also make hardware alterations and develop software apps for different industrial and professional needs, including medicine, media, aviation, education, law, sports, construction, manufacturing, and more. Google Glass is driven to upgrade its functionality and explore more options while striving to meet the needs of business users. Experience in product development and design could be leveraged to better understand consumer needs. Fundamentally it is the same process to cater to either a corporate or individual customer.

One of the major reasons that Google Glass experienced difficulty in penetrating the consumer market was the issue of privacy. A new study from market research firm Toluna found that privacy worries are a major stumbling block for Google Glass, with 72% consumers citing concerns such as the potential for hackers to access private data, the ease with which others could secretly record their actions, and the potential for private actions to become public. Google's pivot to the business user market doesn't mean these privacy issues no longer exist. On the contrary, they become more acute. For example, when used in medical areas, doctors could use Google Glass to record patients' private information or even the entire process of an operation. Once this type of confidential information is compromised, it could lead to a massive data leak. Such incidents could be extremely harmful to users as well as to businesses. If private users have concerns about privacy, businesses could only have more, as they are the caretakers of customers' information, trusted to look after this data. In an era of ever-more stringent data governance (e.g. the General Data Protection Regulation in Europe), the protection and safe storage of data is a critical issue.

As Google Glass is chosen to provide service for these businesses, data protection concerns cannot be avoided, as consumers are indirectly using the service too and they are still very aware of the risks. Google Glass could use this opportunity to save its reputation.

Google was prepared to respond positively to privacy protection-related issues. Google Glass product director Steve Lee pointed out during the Google Developers Conference that privacy issues surrounding the device had been considered a top priority by his team since day one. For example, the prism display is positioned in front of the Glass, so that when it is in use, there is a slight upward angle. This would be the equivalent of lifting up hands to take photos or using voice control to give instructions. No matter how Google tried to respond, the torrent of accusations and complaints resulted in Google Glass not being able to even start addressing this issue for consumers.

Now that Google Glass has entered the business market, the privacy concerns can finally be tackled in two ways. First, it is a good opportunity for Google Glass to defend and prove itself on this issue so that any misunderstanding among consumers can be dealt with. Second, should any of those privacy concerns pose potential risks, Google Glass must face them or business users will raise higher demands regarding privacy protection.

An old Chinese saying goes, "The nail that sticks out gets hammered down." Google Glass has been a landmark product, receiving extraordinary attention and high expectations. Data protection issues pervade the entire wearable tech industry, and they cannot be resolved very quickly. If it is part of the package, once it is successfully dealt with, it would set industrial standards for others to follow.

Since Google Glass has chosen to enter the business market, it will find its way around. If it performs well then it will build confidence for returning to the consumer market. After all, the line between these two markets is indistinct. People are the ultimate users, and building a solid user base is what Google Glass should focus on now.

Some may question whether the focus on the business market could leave Google Glass in the same place that Blackberry once was. Blackberry over-focused on business users and neglected the mass consumer market. The latter backfired into the business market, leading to an awkward situation for Blackberry and the need to regain customer confidence. Google addressed this concern by restructuring its Google Glass project.

On 19th January 2015, Google halted the Google Glass Explorer program, and announced that they would not release any more consumer versions of Google Glass in near future. In another blog article published on Google+, Google said the project would remain, but the head of the unit would be Tony Fadell instead of Ivy Ross. Who is Tony Fadell? Tony Fadell is a consumer product designer and marketing expert. He worked with Calvin Klein, Swatch, Coach, Mattel, Bausch & Lomb, Gap and many other well-known consumer brands. He contributed to the design and creation of Apple's iPod, and he invented one of the star products of the smart home – the Nest thermostat. This staffing arrangement reflects the strategic deployment of Google Glass in the consumer market while remaining strong in the business market. The troublesome history of Google Glass to date may not be a consequence of strategic mistakes, but a reality check for the entire wearable tech industry today. What happened to Google Glass could happen to any other wearable device. Google Glass happened to be the unfortunate guinea pig.

Google CFO Patrick Pichette shed some light on the future of Google Glass. "When our teams aren't able to hit hurdles," said Pichette, "but we think there's still a lot of promise, we might ask them to take a pause and take the time to reset their strategy—as we recently did in the case of Glass."

4. Google Glass applications for business and their commercial prospects

Here we examine a number of major Google Glass applications, in the medical industry and beyond.

(1) Medical industry

Medical applications seem to be the most promising industrial application of Google Glass so far with many successful cases. The hands-free interface and augmented reality functions are the main selling points here. Hands-free technology enables doctors to access real-time information and record procedures while operating on patients. The use of augmented reality has been shown to empower the learning process, resulting in better understanding and stronger practical skills.

Dr. Pierre Theodore, a cardiothoracic surgeon, has been using Google Glass in the operating room at UCSF Medical Center to consult colleagues from other departments or view X-ray images without walking away from the operation table. He is also able to use voice commands to control the device. "There's relatively little shift of attention between seeing the patient in front of you and seeing critical information in your field of vision," Dr. Theodore explains. "I believe it can be and will be revolutionary."

DrChrono, an electronic medical records company based in Mountain View, California, has developed a "wearable health record" app for Google Glass. After registering, doctors can use the company's app to record activities like consultations and medical procedures, with a patient's permission of course. Videos, pictures, and notes are saved to a Cloud-based electronic medical record system powered by Box. In addition to an accessible digital record-keeping system, physicians can easily share information with their patients when requested. Doctors at a major hospital in Boston have become the first in America to use Google Glass for routine work, while surgeons are using it in operations. Medical students at Stanford University are learning surgical methods with Google Glass, using real-time streaming to communicate with teachers. Surgeons at Duke Medical Center have also used Google Glass to record surgical procedures.

Google Glass-based medical procedure models are expected to increase significantly in the future, which will effectively integrate global medical resources to create synergies. In other words, by using Google Glass we will be able to access medical expertise anytime, anywhere. This could make medical diagnosis at a distance possible through remote consultation, or even achieve remote real-time surgical procedures with possible "on-site" supervision. The implication is a Google Glass-enabled global medical system, a new form of medicine delivered through wearables.

(2) Two other industries

Non-medical applications of Google Glass are also emerging. Indiana Technology and Manufacturing Companies (ITAMCO), for example, have released a Google Glass application named MTConnect. Automation World called it "a manufacturing industry standard established for the organized retrieval of process information from numerically controlled machine tools." Even General Electric is trying to bring Google Glass into its manufacturing process.

A specialized wearable solution provider for oil and gas companies is also working on a Google Glass app to enable engineers in the field to have hands-free access to templates, information, and data transmission.

Chris Fleck, vice-president of mobility at Citrix, told PC Pro that Citrix has been working with Google to develop prototype workplace apps for the headgear. "There's nothing production ready yet, but there's a few prototype apps we're working on including video and image capture integration with ShareFile and voice-to-text note capturing, as well as GoToAssist," he said.

Sullivan Solar Power, a San Diego-based solar panel installer, is slipping Google Glass onto the heads of the field technicians who add their panels to the roofs of homes and businesses across Southern California.

These cases prove that Google Glass has tremendous value in potential industrial applications. This may be particularly true in the case of emergency repair work, since with Google Glass, technicians could easily access support from any part of the world. For highly technical installation jobs, workers only need a Google Glass headset to perform perfectly with remote, real-time instruction. More business models will also emerge as Google continues to explore its business applications.

11.4   The Ultimate Aim of Google Glass

When Google Glass Explorer was discontinued amid concerns about privacy, many thought Google was failing in the wearable tech industry. That is because many people do not entirely understand what these global tech giants are trying to achieve. And their understanding of Google Glass was wrong too.

The underlying reason Google entered the wearable tech industry was to control mobile internet data access, and to establish a mobile internet based Big Data search platform. For example, with the acquisition of Nest, Google is not trying to have a presence in the smart home industry, but to likewise establish a Big Data search platform from it.

Building a Big Data platform is the aim of all these efforts in hardware development and acquisitions, as the Big Data platform is the foundation for a mobile internet user base. Google, as we all know, is a search engine company, a Big Data stronghold in itself. Would Google leave its main business behind for a directional change to focus on a physical industry such as wearables or smart homes? Obviously not.

What drives Google to invest so much in developing Google Glass and its on-going refinement and upgrades? We may need to look back at the beginning of this wearable revolution. Who kicked it off first? It was indeed Google through Google Glass. And who fueled the smart home concept? Again, it was Google, with their acquisition of Nest.

While others followed suit to build their businesses in this area, Google seemed to quietly move aside to focus on building a system platform instead. Google wants to be ready to present its dedicated operating systems for wearables and smart home app developers to use. Other players will all need such smart device app systems after conquering technical issues around smart home devices and wearables.

Hopefully, you can now see Google's intention much more clearly. It is not surprising that Google dismissed the entire Glass developing team after their mission was accomplished: to launch the wearable tech revolution. Google decided to develop its own Glass because it was in need of a device to support the construction of a search platform in the era of the mobile internet. Google was able to perform repeated tests with its own products so that it could accumulate knowledge and refine the mobile internet Big Data platform.

In the era of desktop personal computers, user retention was counted in hours or days. We could control the users only with the data platform of those PC terminals. It is a different thing with the mobile internet. User retention is now counted in minutes. It is easier now to go about without a PC, but much harder to go without a mobile phone. We now can see a major difference between mobile internet and wired internet services—the sticking time is much shorter.

This will be further shortened with the emergence of wearable devices. To safeguard its Big Data platform stronghold, Google must focus on user stickiness. It is very clear to Google that ultimate user stickiness in the era of the mobile internet is achieved through coding the operating system of wearable devices.

Google Glass was never the true aim. The commercialization of Glass also seemed half-hearted. Google is more interested in exploring the potential and guiding the direction of the wearable industry, including advances in medicine and healthcare.

Whether a commercially successful product or not, Glass fulfilled its mission to attract global interest, capital, and talent to the nascent business of wearable technology. Among these new wearables, a significant part of them will use the system platform developed by Google. Their users will then build a Big Data empire through their wearable devices. The commercial success of Google Glass should not be measured by sales of the hardware device itself, but rather the advantage it has given Google in the exploitation of Big Data from mobile internet users.

The seeming failure of Glass in the consumer market was in fact only an attempt to test a lab-developed product in a consumer context.

Another contribution brought by Google Glass is the debut of wearable-based VR and AR products.

CHAPTER 12

OCULUS RIFT

O culus initiated a Kickstarter campaign on 1st August 2012 to fund Oculus Rift, a virtual reality (VR) headset. The campaign's tagline promised that Oculus Rift "will change the way you think about gaming forever." Since then it has amazed gamers with its performance at this stage.

The Oculus Rift is a VR headset designed specifically for video games. Its unique performance immediately won the hearts of many. In less than a month, its crowdsourcing campaign racked up USD 2.43 million from more than 9,000 backers looking to fund its development and manufacture.

The two landscape displays in Rift provided a combined 1280×800 resolution, with each eye seeing an image in 640×800. The biggest feature of Rift is the gyroscope-enabled viewing field, which significantly improves the sense of immersion. Oculus Rift headsets could be linked to computers or gaming consoles via DVI, HDMI or micro USB.

On March 26, 2014, Facebook announced its acquisition of Oculus VR for around USD 2 billion, taking an important step into the wearable tech industry. In July of the same year, Facebook announced the additional acquisition of game networking engine RakNet, to make it open source for global game developers. This brought Oculus Rift to a new stage of platform building.

In June 2015, Oculus released its consumer version of the Oculus Rift virtual reality headset at an event in San Francisco. Without establishing its price, Oculus CEO Brendan Iribe announced that Oculus planned to ship the Rift in the first quarter of 2016. Oculus also revealed its partnership with Microsoft: Rift would work with Windows 10 and support all Xbox games.

12.1 Oculus Rift Comes from Gaming but Will Not Stop There

One of the best applications to date for Oculus Rift is gaming. We have seen realistic 3D images in many games before, but they were limited to a screen. But with VR, gamers' sense of immersion can be greatly enhanced. People may have a similar experience of being plugged into another world as in the movie *The Matrix* and its sequels.

Technically speaking, VR headsets are not novel. Research on this topic started over ten years ago. The popularity of Oculus Rift is not only due to its extraordinary gaming experience, but also due to its more acceptable price tag. Before Rift, VR headset devices were in the price range of tens of thousands of dollars. Selling at a few hundred dollars opens up the experience of immersive gaming to the mass market.

A VR experience like this is not quite like the holodeck on the USS Enterprise of *Star Trek,* but it is as close as it gets at present. Almost any gaming type could be enhanced through VR technology, which is the biggest advance in computer gaming since the invention of multiplayer online gaming.

And yet the strength of VR technology is not limited to gaming alone. As mentioned earlier, some companies have already applied it to product design and engineering. Oculus Rift could empower more to do so. Whether the user is a freelance designer or a 3D-printer owner, the technology could be a great asset in the workplace. Virtual tours could be very different when an immersive Google Streetview type of experience is used for exploring a famous building or a town.

How about online shopping? Are you often disappointed with items looking different from their online images, particularly clothing? High-definition 3D models offered by VR technology could be the solution. You could virtually try on each garment. This could be even better than the 360-degree viewing function available with some online shops. This could redefine the online shopping experience by bringing the shop far closer to home than a physical store.

Entertainment, particularly TV, is another area that could be transformed fundamentally. You could imagine what it feels like to be sitting in the driver's seat of a NASCAR or Formula One racing car. How about standing in Santiago Bernabéu Stadium cheering for Real Madrid or in the Hard Rock Stadium watching the Miami Dolphins? The endless potential this technology offers to the family entertainment industry could be revolutionary.

So far there are over 10 different VR headset products on the global market. Apart from Oculus Rift, there are Sony's Project Morpheus, Samsung Gear VR, Microsoft Hololens, HTC's Vive, Google Cardboard, and others. In the coming years, we will witness these devices becoming a part of our lives, changing the way we live and have fun. It may also trigger many business opportunities in all walks of life.

12.2   Why the "Mediocre" Oculus Rift Receives So Much Attention

The development of VR technology has received tremendous attention. It is as trendy as smartwatches and fitness trackers. Well before Oculus, Samsung and Sony both released similar VR products for gaming. In that sense, Rift is not a novel technology. Compared with Microsoft HoloLens and Google Glass, it even looks rather mediocre. And yet as mediocre as Rift is, Oculus is receiving enormous attention at the moment. There are a few contributing factors.

1. A new "sugar daddy" at Facebook

Facebook spent USD 2 billion to buy Oculus VR in 2014. This acquisition rekindled interest in the VR business as a new topic in wearable tech. And it placed Oculus squarely under the spotlight.

In the VR market, Google, Microsoft, Sony, and Samsung all had an existing presence. What made Facebook favor Oculus? Would a social media company be interested in smart devices and want to tap into the field of hardware? Obviously not. We can be certain that Facebook is still mainly interested in social networking rather than hardware sales.

Then why on earth was Mark Zuckerberg willing to spend USD 2 billion to buy a project even before it launched its VR system into the consumer market? We need to understand it from his own words: "VR is the next major computing platform, one that will be a vehicle for communication, shopping, education, and more." Indeed, Zuckerberg was not interested in sales of the device, nor did he have high hopes for VR gaming. Fundamentally he was still interested in social networks and the platforms on which they reside.

I have mentioned before in my book, *Is Facebook's IoT for Real?* that the early stage of Facebook was based on the social lives of students. Facebook provided a platform for young people to find a date quickly. As the majority of the users were college students initially, the lower user segmentation meant the background selection time was short, and so it grew rapidly among American college students.

As investors started to come in, they raised the bar for its business expansion. Moving beyond the college student market, the user base started to grow to related groups and users from other countries. It was more or less an international dating site. How to keep users on the Facebook platform, once they found their dates or when they failed, became the next question. The answer was to find fun things for them to do.

2. The other rich lover

Oculus VR is such a spotlight business and a rising star in wearable tech, with social media giant Facebook on one side, and on the other a close relationship with one of the greatest tech powers, Microsoft.

As extensively covered in the media, the Oculus VR headset is compatible with Windows 10 and the Xbox One. Xbox One controllers were to be sold as a bundle with Oculus Rift. Two messages are important to recognize here:

(1) Microsoft is ambitious regarding the Internet of Things

The announcement that the Oculus VR headset can be used on Windows 10 seems rather insignificant, but in fact it is an important signal from Microsoft— Microsoft has been making moves to be part of the IoT era.

Microsoft was the leader of the PC internet era with its Windows operation system dominating the PC world. However, Microsoft failed to keep its momentum in the mobile internet era. Both Windows Mobile and Windows Phone systems were launched with tremendous ambitions but failed to reap the expected results. Microsoft was outplayed by the hardware geniuses at Apple and the search giant Google. iOS took over the high-end market hands down with a premium operation system, while the open-to-all Android grabbed the remaining market, leaving Microsoft nowhere to go.

Mobile internet is after all the transitional period between PC internet and the Internet of Things. What Microsoft does in the coming IoT era will determine the future of the company. Once every device and facility is connected and digitalized, there will be higher demands for stability, complexity, and security from operating systems. Microsoft has inherited advantages as the operating system leader, in technology, experience, funding, and marketing.

The only issue Microsoft needs to overcome is user-friendliness. In the early PC days, users spent time to learn how to work the operating system. Windows was the center of everything and people worked around it. Mobile internet brought in the user-centered approach led by Apple's iOS. Many decentralized interfaces enabled users to operate the system intuitively. Android goes even further to open the entire platform to developers, making it suitable for their targeted customer segments.

Microsoft didn't adapt to this change successfully, but after years of trying and failing, Microsoft finally came around to recognize this change and made necessary alterations. Windows 10 was released straight after Windows 8, making the point that Microsoft skipped a stage to leap from PC internet directly to the Internet of Things. The lost potential of the mobile internet market may have been a minor setback, but there are greater stakes to play for in the IoT.

One important feature of the IoT is the range and the spread of the smart devices involved. This is very different from the limited hardware types available in the PC internet and mobile internet eras. For example, wearables have many product formats and applicable areas, which pose a big challenge to the agility and adaptability of the supporting operation systems.

Microsoft has made a great effort in this regard. The Microsoft smart tracker, HoloLens VR device and the collaboration with Oculus VR have all had one target in mind, that is to ramp up experience in operating system development for future deployment in the IoT.

(2) The Xbox One hand controller bundle kills two birds with one stone
On the Oculus side, it is not technically possible to quickly develop a stable, functional hand controller to support its VR device. Although Oculus announced its own Oculus Touch wearable control technology, it is fair to say the technology is not yet ready.

As the main market entry point of Oculus devices, gaming customers are a group of users with high requirements in terms of the accuracy and agility for controllers, and also in terms of system response and stability. Gamers are different from "visual" users. The latter only require a smooth viewing experience. But any faults in system stability, control agility and precision could significantly impair the gaming experience in VR.

To perfect its VR product is quite enough for the Oculus team to address. It is not an easy job to turn an idea or a prototype into a consumer product with guaranteed performance. Particularly for a headset designed for long-term wear, any failure in the smallest detail could affect the user's experience.

There are also Rift-compatible games to be developed. Only a few of these games have been released so far. Oculus is a long way off from developing the range and levels of gaming complexity to meet the needs of the huge gaming population brought by Facebook.

It is inevitable to draw the conclusion that finding strong partners in this area might be the best choice for Oculus at the moment. On the other hand, Microsoft hasn't stopped looking for collaborators to further its development of an operating system platform for the IoT. Microsoft is also experienced in the gaming industry with the Xbox, its popular gaming console, which is a handy solution for Oculus. Xbox One was the third generation of Microsoft's home gaming console, a much more developed consumer product than Oculus Rift.

Offering the Xbox One hand controller to Oculus meets multiple objectives, as it immediately boosts the sales of the devices and it brings in the important VR application experience that Microsoft needs while perfecting its operating system. It might not be what Oculus prefers, but it does offer the best available solution. In terms of partners, established companies like Facebook and Microsoft seem to be better choices. Choosing Windows 10 also helps solve many software application problems for Oculus, among which the real-time VR image rendering and self-adaptive performance are the most crucial. VR graphics need to respond accordingly whenever the user turns their head while playing a game. Particularly in the more sensitive visual system, when the eyes don't accommodate the head-mounted screen display precisely, it creates the vergence-accommodation conflict, leading to "virtual reality sickness."

3. As mediocre as it is, Oculus made the first consumer VR product

It is fair to say that the Oculus Rift is not novel, instead it is a rather unexciting product type in the wearable tech field. The phenomenon of Oculus is largely due to the commercialization of this mediocre VR concept.

Google Glass, Microsoft HoloLens, and many other products have one thing in common, that these high-tech gadgets caused a sensation but faced difficulty in becoming mass market consumer products.

Oculus VR is a prime example of such a difficult situation. Without a supporting production chain or consumer awareness, the wearable tech industry is trying to create working businesses from absolutely nothing. Entrepreneurs rely heavily on their own means to keep exploring in this frontier technology industry. Most of their products are not perfected, and yet they still attract huge interest and passion from consumers.

Oculus VR products are exactly so. On the hardware product development level, it is a typical "maker" project with a team that is clearly short of technical expertise and is inexperienced in product commercialization. Just like all other wearable products, it was foreseeable that the first generation of Oculus VR products would have some flaws at the hardware level as well as at the software level, including interactive control, virtual reality display, and other aspects.

To make a concept technology into a consumer product requires entrepreneurship and significant support. This is a new industry and the process is an on-going effort. The Oculus Rift is not the most innovative product concept, but it is an achievement on multiple levels, including system building, hardware development, and technology application. Turning a tech concept into a consumer product is the implementation of all these areas, and that is why Oculus became the center of attention.

But more importantly, the power of wearable technology itself, as I repeatedly emphasize here, will likely replace smartphones to become the next major technological revolution of the world. Virtual reality is a mesmerizing technology that could potentially replace most of the screens today for a screen-free tomorrow. The exploratory efforts made by Oculus and others will attract much attention—sparking both debate and anticipation.

12.3   Oculus Home — the Heart of Oculus

At a pre-E3 (Electronic Entertainment Expo) show, Oculus Vice President Nate Mitchell announced a brand-new VR user interface: Oculus Home. Users would be able to buy Rift-compatible apps from this platform straight away. Mitchell also made it clear that Oculus-exclusive games could only be acquired through the Oculus Home platform, and Oculus apps will be the majority of apps available on this open platform. Mitchell also said that Oculus had opened its platform to Gear VR. When the Oculus Rift app went online, all apps from Oculus Home and apps from Oculus store on Gear VR were supported by the same platform.

From Mitchell's presentation, it was noticeable that the focus of Oculus was not the VR device itself, but the construction of an Apple Store-like developer platform. The only difference is that Oculus Home is solely focused on the gaming developer segment at the moment. The company also announced its investment of USD 10 million to fund developers in creating more Oculus tools and programs for VR.

It will take the Oculus game developing team a long time to offer products for gamers on Facebook. The need is not likely to be met there, much less for users from other platforms. The only way out is to follow Apple and focus on building a solid product and a reliable system. These will form the basis of a successful developer platform to invite global gaming developers. They could expand into more application fields of Oculus VR and present the users with more VR apps and experiences. As Mao said, we need to unite all forces that can be united to win this revolution.

12.4 Gaming Is Just the Beginning

Gaming is just the beginning for Oculus. There is one topic we must address: pornography. Whether you like it or not, if Oculus wants to be commercially successful, it needs to work with the porn industry. The largest porn website, Xvideos, receives over 4.4 billion visits each month. That is three times the number of visits to CNN or ESPN websites, and twice that of Reddit. Other pornography sites like YouPorn, Tube8, and Pornhub also achieve much higher clicks than Youku, Tudou, BBC or Tmall.

In terms of the average time per page, there is also significant difference between porn and non-porn sites. Well known tech sites such as Engadget or ExtremeTech achieve on average 3–6 minutes of visit duration, but porn sites can achieve between 15–20 minutes per average visit.

This data reveals the true interests of many people. Therefore, when Oculus VR meets pornography, it provides a milder way for users to view shows with total privacy protection. On-demand payment options for streaming services could be made available too. This technology also makes meeting potential dates a totally different experience. A VR interview could be as real as meeting in person.

It is worth questioning the possibility of certain sex businesses being replaced by the Oculus Rift. Users could select services from an online store and meet the person through VR to talk about their options.

PART 3

WEARABLE BUSINESS MODELS — A GLIMPSE INTO THE FUTURE OF BUSINESS

Wearable technology is still in its infancy. Most business models still have a focus on the development and sales of hardware and accessories. As the industry develops, an ecosystem will establish itself, providing support for many potential business formats. Free from reliance on hardware sales as the sole profit-making channel, Big Data-based wearables will trigger new business opportunities in healthcare, tourism, education, gaming, fitness, advertising, public life, and many other areas. This part of the book examines the potential developments in a range of sectors. As we look ahead, we can imagine the potential for radically different business models—models for which businesses today should be conscious of, and the potential gaps in the market that can be exploited.

CHAPTER 13

WEARABLE MEDICAL TECHNOLOGY

The potential for wearable technology in hospitals, health-related areas, and other medical applications is immense. This may include the use of sensors in providing care, using the latest wireless and Bluetooth technologies to collect real-time measurements from different parts of the body to enable better treatment, whether preventative or when intervention is needed. Medical care could be transformed by the use of different sensors, with machine learning-generated algorithms supporting medical staff in the process of diagnosis and treatment. Wearable technology may allow for early intervention before a patient is even aware of their ailment. Wearables can also assist in delivering healthy lifestyles—in terms of diet or exercise—and provide data to medical staff on users' activities and other measurements of interest.

The use of systems approaches, based on huge new platforms, is needed to support the use of wearable technology in this area. There are a number of key components of a system that need to be in place to ensure the health benefits of wearables are fully captured. To date, most wearables—including smartwatches, wristbands, clothing, and footwear—are only used to monitor user body readings for the purpose of fitness management. This could help address a number of public health issues, from obesity to cardiovascular disease. However, this technology has not yet been integrated fully into medicine. As we look ahead, there is a plethora of potential applications of wearables in the medical area—fitness management is but an early step in tackling the health concerns facing the planet in the current century.

Of particular interest for the wearables industry is the potential for mobile health, or mHealth. This is discussed in the next section.

13.1   The Optimal Hardware Choice for mHealth

A number of different commentators watching the wearables sector have identified the potential value of such technology in medical applications. According to a study by Transparency Market Research, medicine is the most promising application field of wearables, with prospects exceeding those in the fitness and entertainment areas.

Wearables could revolutionize the medical device industry. We have seen a number of devices shrink significantly in size—from large bulky machines to portable devices, and finally to a scale that can be worn with little or no discomfort. Using such devices enables real-time monitoring of a number of metrics important for health at any location. Wearables can currently give high quality data on a range of significant attributes—from blood sugar to blood pressure, oxygen saturation in the blood to body temperature, and respiration to movement. The list of potential measurements that can be obtained from wearables is growing at a rapid pace.

Beyond health measurement, wearables can even be used for direct medical treatment. iMedia Research predicted that the market size of medical wearable devices in China would reach RMB 1.2 billion (USD 171 million) in 2015, and RMB 4.77 billion (USD 681 million) in 2017, with a compound annual growth rate of 60%.

Wearables are well-suited to be the main hardware supporting mHealth. This is for a number of reasons, notably:

1. User base development

Before the arrival of the mobile internet or wearable technology was known, there were blood pressure monitors and blood glucose monitors sold on the market. In China, these devices are sold by the millions every year. The demand for this kind of product is still growing. With more advanced technology, the functionality of wearables will improve and so too will the ability of wearables to actively help manage the overall health of the user. This will encourage the entire user base of this market to use more and more sophisticated and comfortable wearable medical devices.

2. A truly hands-free tool

Wearables will replace smartphones because they have the advantage of being hands-free, offering new input and output terminals for the mobile internet network. Instead of requiring the use of typing, set gestures (e.g. raising an eyebrow) or voice activation, the input of wearables could be completed while they are worn, like smartwatches on one's wrist. The input data would consist of body signals such as heart rate, blood sugar, or even brain waves.

3. Body monitoring around the clock

As long as the user has a functioning battery, the wearables could record data from embedded sensors and process the data for analysis. Most wearables today, whether they are glasses, watches, wrist bands or such, could be easily taken off and left aside. But the future wearables may employ a more integrated format to become a subtle addition to the human body. Sensors could be implanted into all of our clothing, our footwear, and even our bodies.

4. The chemistry is developing

The concept of mHealth has been around for quite a while but has not yet been fully implemented. Most mobile health services today work through the simple connection between an app and a smartphone. This, it can be argued, is not true mHealth, as the key value in the data has not been mined for use.

Arguably, the chemistry between the component parts of mHealth is still developing. The integration between sensor technology, the IoT, 4G or even

5G connectivity, Cloud computing, Big Data analysis, and other components behind mHealth intervention, is only just starting to become technically feasible. These combinations will multiply—offering the potential for radical innovation. Wearable technology for mHealth is but one outcome of the interweaving of technology, people, and data systems.

When deep integration occurs between these technologies, data mining will provide many answers. Targeted health advice for a range of conditions will be based on algorithms fed by smart wristbands tracking and storing data on different health measures. This will be facilitated by data processing through an established ecosystem based on the internet, using Cloud computing and mobile health platforms.

Internet connectivity and access will no longer be a limiting factor. It will become embedded in all sorts of items we see and touch.

In some ways, the combination of sensors added to things we wear opens up a number of new potential functions going beyond those seen and used in smartphones. The collection and analysis of data from humans will give physicians access to better knowledge to support healthcare. The personalization of healthcare, combined with improved access to physicians using technology, will radically redefine healthcare systems. It will potentially open doors to improved self-monitoring and treatment of medical conditions and give patients agency in the control of their own health data. Physicians and patients alike will be supported by better data hosted on platforms designed to take and process data from wearables and combine it with other data (including both health—e.g. prescriptions—and non-health data—e.g. climate).

The need for physician-led intervention in common diseases could even be significantly reduced through the use of AI-enabled diagnostics. This would clearly need to be appropriately regulated, with stringent standards in place for medical wearables and their systems. However, one can foresee a time when people may be assisted medically more by machines than human doctors.

13.2   Key Opportunities in the Wearable Device Market

Starting from the current market for wearables and gazing into the crystal ball, I believe that medical wearables have the best future market prospects of all wearables in existence today. The segment with the largest potential includes those with chronic diseases, who exhibit rigid demand. The truth of this can be seen from a number of angles:

1. The sheer size of the population suffering from chronic diseases
The World Health Organization (WHO) issued a report on January 19, 2015, stating that among all causes of death, non-communicable diseases (NCD) including cancer, chronic respiratory diseases, cardiovascular diseases, and diabetes are still the main killers. Many premature deaths from these causes are preventable.

According to WHO statistics, in 2012 over 38 million people around the globe were killed by NCDs, including 8.6 million in China. Among all NCD deaths in China, 39% of the men and 31.9% of the women were premature deaths, totaling over 3 million.

Providing crucial insights into the incidence of NCDs in China between 2003 and 2013, the China Health and Family Planning Statistics Yearbook 2014 drew on extensive analysis to show that:

- The average incidence of diabetes rose by a multiple of 7 during the decade, with urban cases increasing by a factor of 3 and rural cases by as much as 10x. Apparently rural China was more severely hit by increases in NCD over that period due to radical changes in diet, living habits, and environment.
- The average incidence of hypertension rose by a multiple of 6 over the ten-year period, with incidence in cities rising by a factor of 3 and rural incidence by as much as 8.
- NCDs impacted 12.8% of the entire population in 1998, rising to 15.7% by 2008. Such a trend is likely to remain. Meanwhile, more NCDs are occurring among younger people. Statistics show that more than 65% of the Chinese labor force has NCDs. 69% of hypertension cases and 65% of diabetes cases

occurred during the working life of people. NCD deaths account for 85% of all mortality in China.

- Spending on treatment for NCDs accounts for nearly 70% of health expenditures. The World Bank forecasts that a 1% decrease in mortality from cardiovascular diseases in China could create economic benefits reaching USD 10.7 trillion.

The above summary clearly shows that NCD management and care could be a huge market for smart wearable devices. With relatively high entry barriers and rigid demand, the potential for profitability is significant. The early detection of chronic disease, coupled with reductions in premature deaths through preventative measures, should be a priority. Using advancements in wearable technology to tackle these massive societal problems will allow for the realization of just a fraction of the ultimate value of the technology itself.

2. Existing consumer awareness and wearable retention rates
For most existing wearables, it takes time for users to get in the habit of wearing them. When a company brings a new technology to the market, building customer awareness can involve incredibly expensive marketing campaigns. In addition, fitness wearables have needed to evolve to meet changing consumer preferences to avoid becoming outdated. The use of interactive social media tools to gamify the use of wearables can help change behavior. Time is needed before users fully integrate the novel device into their lives—and become comfortable with the concepts of tracking and measuring their body's performance. Many tech start-ups struggle to meet the costs associated with building and maintaining their customer base. Deep pockets are needed to survive in this market.

The situation is quite different for medical wearables, particularly in their use for the management of chronic diseases. Before wearables became popularized, devices like electronic blood pressure meters and blood glucose monitors were being used to improve the daily lives of those with certain medical conditions. Smart wearable blood pressure meters or smart blood glucose monitors can offer new functionality, but they are based on similar concepts of self-monitoring. They may look different than the old devices, but they may also contain vastly improved technology—including the potential for linking to other devices and access to one's own data.

Patients (or consumers) understand what blood pressure monitors do, so it is far easier for users to adopt these than wearables where the functionality may be less familiar or understood. Such understanding makes the uptake of the technology more likely and probably allows for quicker diffusion of the technology in the market place. For companies, the fact that patient or consumer awareness is high shortens the "go-to-market" phase and provides significant cost savings.

3. *High stickiness of NCD patients to their medical wearables*

An outstanding feature of NCD patients is that their demand is highly focused on the technology's quality, rather than style or customer service. They are unlikely to abandon a device when the novelty factor wears off or because they dislike some element of its appearance. As long as the device delivers the required functionality, NCD patients will be satisfied.

For example, patients with hypertension need to measure their blood pressure and take medication regularly. Their need is for a monitor that can measure blood pressure accurately. Wearable device developers therefore focus largely on making an accurate and easy-to-use device that attaches to the patient's body to allow automatic blood pressure monitoring. The device is also designed to transmit data to the patient's smartphone, which can be viewed in an app. The app may also provide diet and lifestyle advice to help the patient maintain a stable blood pressure. The data could also be shared with clinicians in order to reduce the need for consultation or to enable early intervention as needed.

The elderly can also benefit, particularly from smartwatches and smart clothing which measure heart rate, temperature, posture, movement, and other factors. Dr. James Amor, a research fellow at the University of Warwick, says activity monitoring allows families and caretakers to see the elderly person's health and routine. Through the deployment of wearables, community healthcare services could add to digital health records for all residents, which would help to better understand the impact of chronic diseases in the community. The data collected could also be used to assist in disease management and in medical research.

NCD patients will likely become reliable consumers of medical wearables. They are also the users who truly need such technology. In the context of an aging society, with significant burdens from chronic disease, medical wearables may open doorways to new solutions and so improve lives for patients and reduce costs for healthcare providers.

4. Significant lowering of medical costs

Currently NCD patients require frequent checks and reviews, as part of a long-term treatment plan including medication, in order to control the progression of the disease. Patients also need to pay regular visits to doctors, with significant costs in terms of both time and money.

Take diabetes, for example.

On World Diabetes Day in 2012, the Chinese Diabetes Society (CDS) and the International Diabetes Federation (IDF) issued a joint report on the social and economic impact of diabetes in China. This showed:

1 Direct medical costs related to diabetes account for 13% of the nation's entire medical expenses, reaching RMB 173.4 billion (USD 25 billion). The average cost of care for diabetic patients is three to four times higher than other patients (including the need for hospitalization and outpatient treatment).

2 Medical expenditures on diabetic patients are nine times higher than non-diabetic patients of the same age and gender. Patients who have had diabetes for 10 years or more cost 460% more than those with a history of diabetes between one and two years. An astonishing 22% of the entire household income of diabetic patients is used on medical treatment related to their diabetes.

According to the China Health and Family Planning Statistics Yearbook of 2014, there were over 98 million people with diabetes in China, making the disease a significant public health issue. A great deal of effort has been spent to find a good strategy to tackle this issue, with the integrated use of eHealth, mHealth and wearable technology being proposed as a possible way to reduce the burden of the disease.

During the Boao Forum for Asia Annual Conference 2015, in the Smart Healthcare and Wearables session, Simon Segars, CEO of ARM, spoke of the major potential value of wearables in terms of lowering costs for medical care. Patients from remote areas could be supported through remote medical technology with the help of medical wearables, saving the costs of travel and hospital examinations. Regular medical checkups could be accomplished just using the data from wearables, to keep better track of the patient's condition. Daily medication, diet management, and other health management functions could prevent the progression of the disease and help reduce overall medical costs.

5. The way ahead for medical wearables

How can medical wearable devices best exploit the opportunities presented by the market demand and the favorable policies the Chinese government has enacted in its healthcare reform? The R&D and manufacturing divisions of medical wearables companies clearly need to play key roles. They need to have clear targets and build the capacity of their staff and operations.

Most NCDs have some common features. They exhibit certain patterns and can be managed through regular medication. Without medication the condition could worsen, for example through increased blood pressure, irregular blood glucose, or abnormal heart rhythms.

From the functionality provided by wearables already on the market, we can see the power of these devices in terms of their ability to collect different types of data. However, having excessive and diverse functionality has been shown to limit R&D capacity, resulting in inaccurate and fragmented data tracking. Such products do not offer much value for users.

Instead if only one function or a consistent set of functions is featured on a device, for example to aid NCD patients in monitoring their conditions (i.e. blood pressure, blood glucose level, and heart rate), then better quality data may result. That in turn has significant value both for the patient and for use in Big Data analysis. By focusing on limited functionality, the manufacturer of such a device is also less restricted by the under-developed production chain in terms of technology, performance, and user experience. Concentrated developmental efforts are most efficient to ensure product reliability. In other words, less is more. Building a focused device to perfection is the best way to become established in a market.

For example, if a blood pressure meter is targeted towards older users, it would be best to design a wearable around older users' needs and living habits, making the device easy to operate. If the target users are middle-aged professionals the design of the device could follow the Apple Watch, being fashionable and polished. For older users, on top of accuracy in data measurement, other functions that assist in the use of the device would be very useful too. Such functions may include voice command and text-to-voice communication of the monitored data, reminders and alerts of abnormal readings, as well as giving caregivers access to the data.

Many years have passed since the first consumer-targeted electronic blood pressure meter appeared on the market. Technology has advanced and the functionality offered by point-to-point connectivity is ready to be exploited in the management of NCDs. Smart blood pressure monitors offer much potential in the management of different conditions, including self-management of the condition and realizing when medical intervention may be needed. Smart devices can help people access care when they really need it, improving quality of life and lowering costs to society and healthcare providers.

13.3   Customer Segmentation

According to a 2014 study on wearables by the market research institute GfK, one-third of consumers abandoned the use of their devices within six months of purchase. Wired Magazine also reports that more than half of US consumers who have owned an activity tracker no longer use it. A third of them took less than six months from unboxing the device to putting it in a drawer or giving it to a relative.

There are multiple reasons contributing to the low retention rate of wearable devices. Contributing factors may include unappealing design or poor operability. Nevertheless, I believe that the fundamental issue is the lack of rigid demand for the functionality offered. If user needs are not appropriately identified, then retention may suffer and so may the longer-term viability of the product in question.

It did not take long before users found out that many heart rate monitors and pedometers are not very accurate in their measurements. This lack of accuracy may lead to the alienation of consumers and may be the reason products with health management, fitness/well-being, or social/entertainment functions all fail to have a long-lasting impact and are often quickly abandoned.

It is critical that wearables no longer try to be the all-purpose products that they seemed to be when they first appeared on the market. Instead, segmentation of the market is the key to developing products with unique functionalities for specific groups of people. This is particularly true in the area of medical wearables. With precise segmentation, users could become strongly attracted to the products. It is then possible to dominate the market and to become an everyday item in the daily life of those users. These elements form the basis of the general direction that the future medical wearable market needs to take.

Based on this segmentation, marketing plans could be more specifically targeted for various products, including trackers, smartwatches, smart glasses and others. It is important to segment the market according to user groups. For example, infants and toddlers, children, women, elderly or disabled people all present different requirements for the wearables they need. These are discussed below in turn.

1. Wearables for infants and toddlers
As this user group is very young and requires special care around the clock, it is important that wearables for them meet stringent safety standards.

Wearables can offer a range of functions to support the care of this group, including monitoring sleep quality, movement, body temperature, heart rate, and other health indicators. Parents can access data and view analysis from a PC or a smartphone. In the case of accidents, such as an infant crawling or falling out of their bed, alerts could be issued immediately to their caregivers.

Quite a number of wearable devices targeting infants and toddlers have been developed recently. They may be a great help to new parents in the first years of parenting. Among the technology in this category is a smart ankle bracelet from the company Sproutling. This monitors the baby's heart rate, sleep position, and whether they are asleep.

The product is made up of three modules: sensors, a band, and a smartphone app. The band is made of soft medical-grade silicone with a lovely red heart design in the middle, embedded batteries, and four sensors. Sproutling also established a health database for babies under 12 months old. Parents can configure the baby's profile by adding their age, body weight, body height and other measurements to connect to the device through a smartphone. When irregular results are detected, an alert is given to the parents through the app. This may give reassurance to parents, particularly after the child has been moved to their own room.

Similar products now available on the market include a smart sock developed by Owlet, smart baby bodysuits by Mimo and Exmovere, and smart diapers by Pixie Scientific. So far in China there are no independently developed infant wearables. There is one smartwatch product called B-smart by Babytree which specifically targets pregnant women. It works on the Android system with apps to monitor body weight, fetal movement, contractions and other data during pregnancy. Tracking fetal health is very complicated and it cannot be denied that this market offers much promise. It is possible that Babytree may diversify into wearable baby monitors as their business develops, but such plans have not yet been revealed.

Prenatal and postnatal care for mothers and babies is focused on a very unique user group. If wearables are safe and accurate in tracking and monitoring data, and are appropriately supported by data analytics and storage, then the market potential could be incredible, and this may well become a source of rigid demand in the future.

Wearable companies could extend their baby products further to target young parents who are in need of parenting help. Programs and courses could be offered as value-added services to these customers. Companies could also look into the possibility of upgrading products according to children's growth needs. This would not only increase user retention but also provide on-going product support throughout childhood.

All in all, the top four needs for infant wearables to be successful are safety, comfort, accuracy, and responsiveness. Those enterprises wishing to enter this market need to have quality hardware as well as a well-established service platform. Those traditionally in the baby product industry in China have a strong advantage in their knowledge of the market and existing customer base. It is possible that these companies will launch wearable products for babies first, in order to take a significant portion of the pie offered by this market segment.

2. Wearables for children

Statistics show up to 200,000 children go missing in China every year, and only 0.1% of these were recovered. A missing child has significant impacts on any family. As a consequence, child safety is now an increasingly important public security matter in China.

Most children's wearables on the market today have very simple functions, mainly GPS and location tracking. Undeniably, this is another major battlefield for wearables with strong market demand. Most of the existing child safety wearables use a 3-in-1 operations model: hardware + software + Cloud. Wearable manufacturers have developed not only the hardware device, but also the mobile app and the data analytics platform. A well-designed combination of these can provide better user experience and more significant returns.

At the end of October 2013, Qihoo 360 launched a child safety smart band at the price of RMB 199 (USD 28). The Qihoo 360 offers a real-time child locator, safety zone alert, voice intercom, and other functions. Globally, there are products like FiLIP smart locator and communicator. FiLIP uses GPS, WiFi, and the cellular network to locate the child's position. Parents are able to monitor their child's location from their smartphones, and the wearable can be used for voice communication.

Another Chinese company Eachpal released a smart band for children called SmartUFO, which uses a WiFi locator. By scanning WiFi hot spots nearby to acquire MAC addresses, SmartUFO is able to locate the device within 20–100 meters. The WiFi locator uses only one-twentieth the power consumed by a GPS locator, therefore the device's standby time is been stretched to two weeks.

Wearables for child safety reached the market in 2013, with more devices of this type appearing in the years since. Beyond the products mentioned above, there are the hereO GPS watch for kids (targeting 3–12-year-olds), the Tinitell wearable phone for children, the LeapBand activity tracker, and the LG KizON. All these wearables include a locator function, with the same problem: the device is useless when taken off the child. Children could take off the device or lose it by accident, or kidnappers could deliberately remove and discard the device.

Companies or entrepreneurs wishing to enter this area of wearable tech must address this critical issue. They also need to look at extending battery life, making location functions more accurate, reducing radiation output, and other related aspects to help create successful wearable products for children.

3. Wearables for the elderly

The aging population—and their associated healthcare issues—is a global problem. The cost of care is rising. To address this, wearable developers could look into developing a system of community care based on Big Data analytics to support care in a person's own home.

China is a country with a strong cultural focus on honoring and caring for the elderly. Looking ahead, smart homecare solutions may become mainstream solutions to caring for the elderly. According to statistics, 90% of older people would prefer to stay in their own home, with only 10% choosing assisted-living facilities. Given that preference, extending care service to those living in their own homes is critical.

The rise of people with Alzheimer's or dementia adds significant pressure to care systems. People with this disease partially lose their ability to look after themselves, struggling to remember things, losing spatial awareness, and easily becoming lost. For this particular market, wearables must have a locator function on top of sensors for heart rate, breathing, and other health indicators.

This type of device is still in the early stages of development. We have not yet seen any major products appearing on the market that specifically target this segment. The main formats of potential wearables are shoes, mobile phones, and accessories. CMA800BK, a credit card size Comfort Zone product recommended by the Alzheimer's Association, weighs only 50 grams. Placed in the pocket of the patient, caretakers could access a patient's precise location at one click. When the patient steps outside an established "Comfort Zone," caregivers receive text alerts immediately.

Utilizing Qualcomm's inGeo platform, an embedded GPS chip, and an internet connection, it can be remotely controlled. The current sales price of the product is USD 99.99, but there is also a monthly service charge of USD 14.99. If the elderly patient is able to carry a smartphone, using a Sprint smartphone with built-in Comfort Zone functions achieve the same functionality with a monthly charge of only USD 9.99.

Many of those with Alzheimer's disease have the tendency to wander. They can be stubborn and refuse to accept change, including the use of new wearables. It would normally take a while before the patient is used to wearing or carrying the device with them. To address this issue, GTX and the shoe maker Aetrex have jointly developed a GPS-based shoe product, so the locator can be worn without the person noticing.

GPS-enabled shoes are no different in appearance than ordinary shoes. The embedded GPS sensor allows caretakers to locate the patients through smartphones or computers. The same "Comfort Zone" alert function is available too.

We have seen some companies in China using smartwatches to target health and care services for the elderly. Using the smartwatch, a connection is made between the user, the hospital, and the family. Apart from daily health monitoring and reminders, it could be used during an emergency to call for medical attention.

4. Wearables for those who are overweight and obese
According to the WHO, by May 2014 there were over 2.1 billion people on the planet who are clinically obese or overweight, accounting for 30% of the entire population. This problem has brought with it significant costs for both rich and poor countries.

The obese and overweight population in China stands at 46 million, the second highest in the world. Zeng Liang, VP of Baidu, pointed out recently that Big Data reveals the top search term for their 290 million female internet users is "diet."

Losing weight is indeed a rising concern as people start to enjoy better material lives in China. Dietary changes in China have increased the risk of obesity in the younger population. In the past three decades, the overweight population of boys and girls in developed countries has reached 24% and 23%, respectively, while in developing countries 13% of children are either obese or overweight—and these numbers are still rising.

WHO's latest report also shows that over 3.4 million adults die from cardiovascular disease, cancer, diabetes and arthritis caused by obesity. It is clear that obesity is a matter of life and death. Yet so far, no country has managed to find an effective solution.

In solving the obesity problem, fad diets or liposuction are not permanent solutions. Only through regular exercise, healthier diets, and better living habits can this issue be properly addressed.

The true issue, when targeting this segment of the market, is how to drive this group of people to change their behavior in order to form healthy habits.

Wearables could help in bringing about behavior change. They've been shown to be one of the most popular product types in the diet and fitness market. Wearables have four unique strengths.

(1) Some wearables can be worn 24 hours a day.
This contrasts with smartphones, which are usually turned off or moved further away from users during the night.
(2) Wearables can monitor real-time health data around the clock.
This is the greatest benefit of wearable devices. The data generated could be applied in many ways to impact daily life, particularly in the field of medicine. Wearables could form the basis of a new era of disease prevention and management.
(3) Social communities formed around wearables and the connection with health insurance may encourage users to adopt them.

If wearables are the social norm or health insurance costs are reduced through their use, they may encourage more people to exercise regularly and have healthy lifestyles. The current trajectory of the market for fitness devices and apps is one of high growth—as more people wear them so they will increasingly become the "norm."

As I have repeatedly stressed, the healthcare industry is one of the first growth points in the wearable market. Based on market segmentation, it is likely that eager dieters will be another stronghold of demand for wearable technology.

It is clear that both the market background and the strengths of wearables have enabled wearable technology to be well-exploited in the fitness market. I therefore would like to encourage investors and entrepreneurs to consider these segments as potential entry points into this very promising market.

5. *Wearables for people with disabilities*

Many people around the world have devoted their lives to helping those with disabilities live more fulfilling and socially integrated lives. Wearables offer great potential to foster integration and reduce barriers for people with disabilities. It is very likely that through specially developed wearables, people with disabilities could live more independent lives. There is great market potential and room for development in this area, for example: using an exoskeleton to help paralyzed patients stand, using special glasses to regain vision for the blind, or helping the deaf to hear with advanced equipment and technology. Many science and technology firms have been developing and commercializing these products. Some of these are discussed below.

(1) Artificial eyes

A biotech start-up has developed a novel artificial eye product. It uses a system called EYE (Enhance Your Eye) to regain sight for the blind. The company uses 3D bioprinting technology to make artificial human organs. They have so far produced ears, blood vessels, and kidneys. As the head of the company explains, due to the complexity of the human eye, it is not easy to print an eyeball.

To date the company has managed to offer three different models of the EYE system. "Heal" is the standard version with an electronic iris. The "Enhance" model adds an electronic retina and camera filters (retro, mono, and others). The version with the highest specifications is called "Advance," which is even equipped with WiFi. Eyes are now electronic devices just like smartphones.

To use EYE, the patient needs to remove the original eyeball and implant what is known as "the Deck," which connects the eye to the brain.

The developers explain that the product will not be available on the market until 2027, therefore no actual photos are available yet.

(2) Tension control devices for autistic children

Autistic children could face tremendous challenges when under pressure and can struggle to voice their needs. Parents and teachers must be extra careful not to stress these children.

Wearable manufacturers have so far launched two products, Neumitra and Affectiva, to address this problem for autistic children by monitoring the physiological reactions of the user. These devices can also be used for medical purposes by monitoring post-trauma anxiety. These smart bands could benefit thousands in the autistic population, making it easier for their caregivers to track their tension levels and provide support.

Some institutions have started to test the Affectiva smart band. The Autism Society of Ohio claims that teachers have been giving out these smart bands to monitor autistic children's behavior. Such data could be analyzed to identify the most relaxing time for them.

(3) Mind-controlled wheelchairs

Emotiv, a Silicon Valley start-up, launched a Designathon to invite its community to use the company's neurotechnology to build new applications. One of the applications was a mind-controlled wheelchair for people with motor difficulties.

Emotiv has developed a headset that can translate mental commands into electronic signals that control the movement of the wheelchair. "Brain-controlled wheelchairs are low-hanging fruit—it's 100 percent the future for wearables," said Redg Snodgrass, founder of media and event startup Wearable World.

The idea of a mind-controlled wheelchair was first proposed by Albert Wong, a Malaysian law graduate with Duchenne muscular dystrophy. His family contacted Emotiv and requested a system that could enable Wong to control his wheelchair and communicate better using a combination of mental commands, facial expressions, and head movement. Although unfortunately Wong passed away soon after this, Emotiv continues to work closely with people with disabilities, particularly with those paralyzed from the neck down.[3]

(4) Emotiv Insight Brainwear

This wearable brainwave monitoring device is the result of the Emotiv Insight Brainwear development project, jointly conducted by Accenture and Philips. It could offer significant benefits for those with motor neuron disease by enabling users to control devices and environment using just their thoughts.

Emotiv Insight Brainwear scans EEG brain signals. It can be connected to a tablet app, enabling users to send brain commands to control Philips products, including its Lifeline Medical Alert Service, SmartTV, and Hue wireless lighting systems.

(5) Motion Savvy

MotionSavvy is a company that emerged from Leap Motion accelerator Axlr8r. What sets this company apart is that all nine members of its team, including CEO and founder Ryan Hait-Campbell, are all deaf. Hait-Campbell learned how to speak through machine assistance.

Ryan wanted to develop an effective and affordable device to help other deaf people communicate. Several years ago, he met software engineer Alex Opalka, then Jordan Stemper and Wade Kellard, with whom he jointly founded MotionSavvy.

MotionSavvy's product is UNI, which includes hardware and software components. The hardware part is a tablet similar in size to an iPad mini. It has an embedded Leap Motion sensor. When the user wants to talk, the Leap Motion sensor captures their sign language and translates this into text displayed on the screen, with a text-to-voice option to read out the text.

In 2017, MotionSavvy released UNI. MotionSavvy had been recording sign language for the device for a year prior to this. Hait-Campbell plans to make MotionSavvy an app system like Google Translate, recording sign language at one end and translating it into spoken language at the other end.

6. *Wearables for professional sports*

Distinct from general fitness wearables, smart devices for professional sports require more precise measurement of heart rate, breathing, and other metrics to monitor an athlete's performance in terms of speed, distance, and endurance. Enhanced data processing software can translate this data to allow sports doctors to understand the athlete's body condition and inform bespoke training or recovery programs. Coaches could also use the information to understand their teams in order to pick the players best suited for the match.

These products have not received much attention from non-professionals. The smart wearable products include clothing and equipment specially designed for sailing, mountain climbing, boxing, and other sports. These products are different from other fitness products because they are able to deliver targeted professional training programs. They can also be used to identify mistakes during training and so to correct these.

When Zlatan Ibrahimović playing for Paris Saint-Germain took off his top after a match, it revealed a dark garment shaped like a sports bra underneath which sparked curiosity among fans. In fact, this was a smart sports vest made by GPSports which was used for tracking a player's body condition and movements in real time.

Apart from Paris Saint-Germain, Real Madrid, Chelsea, and many other top football teams in Europe have also started to use GPSports products. These so-called "Man Bras" are kept well-hidden most of the time.

Other similar wearables include Bro, a rugby top with an embedded data recording device. This product is also a GPS tracker, to help coaches and doctors better understand player movements. When body condition or fitness levels are reduced, coaches can see direct visual data analysis on an iPad or PC, which could enable them to choose players in better condition to play.

CHAPTER 14

TOURISM

14.1   Continued Growth of Online Travel Services in China

As China has become wealthier and mobile internet expanded its coverage, the Chinese have shown an increased desire for travel experiences. There has been exponential growth in the travel market in China, while the growth of mobile internet has made it possible to access online travel services anywhere at any time. Travel and tourism are responding as an industry to this shift in the way consumers want to access travel and tourism products.

By June 2014, the internet user population in China had reached 632 million, with an internet penetration rate of up to 30%[4]. Mobile internet users numbered 527 million.

In 2013, online bookings for travel were worth RMB 218.12 billion (USD 31.16 billion) in China, with a year-on-year growth rate of 27.7%. Online penetration for this market was 7.7%. The growth of the online travel market is largely determined by the actions of related businesses, including the decision to offer flight and hotel bookings. The shift to online travel booking has led to extraordinary growth in some sectors, with new business models emerging. For example, the markets for online short-term holiday rentals, online car rentals, and online taxi booking services have seen significant increases.

We can anticipate that travel and tourism will quickly embrace "smart" concepts. Building on the increased use of wearable technology, future business models for the travel and tourism industry may take on new forms.

14.2 Wearables: Make Smart Travel Personal

A truly memorable journey will often require a great deal of time devoted to the planning phase. We can consider the case of budget travel, and unpack some of the key factors that need to be considered in preparing for the trip. These include the following: the overall itinerary, necessary transport arrangements (tickets or rentals), tour booking, checking the weather, researching the destination to plan where to visit, budgeting, arranging documents (e.g. passports and visas), packing suitcases (including clothing for different weather conditions, cameras, chargers, battery packs, and more), and learning some of the language if traveling to a foreign destination.

Planning a great holiday can be a lot of hard work. How might the tourism industry look in a future with increased use of wearable technology? Picture this in your mind. One day, you are slumped on the couch trying to find a new TV series to binge, but you can't find anything interesting. You're wearing a garment with subtle sensors that pick up your emotions. Combining this with your social media activity, your bank balance, and diary in an algorithm based on your preferences, the smart system is able to suggest that perhaps it's time to recharge your batteries with a vacation. The screen offers you a range of holidays based on your tastes—and you're one click away from being whisked away to a beach resort on a tropical island drinking cocktails. The choice remains yours—will you go or not?

Wearable technology has already brought profound changes to the healthcare, gaming, and fitness sectors. Such changes will inevitably extend to more areas, including travel and tourism. The internet has made planning trips easier through better information. Travel in the wearable age will be easier still. The devices would filter information for you, giving you access to a better travel experience.

How could wearables deal with some of the most common issues in travel and tourism? The next section discusses some opportunities.

1. Trouble-free solutions
(1) Identity recognition
The need to prove your identity is becoming more and more common. Passports, ID cards, and bank cards are indispensable these days when traveling. You need ID documents and cards with various levels of security. Technology has advanced rapidly with facial recognition and other security measures that are implemented in airports and hotels. The use of biometrics is becoming more common—using fingerprints or other metrics for identification. Encrypted biometrics will likely be mainstreamed as the preferred way of verifying a person's identity to give access to social media accounts, smart devices, hotel rooms, and more.

To make this method absolutely safe and secure, wearables are the ultimate solution. Why is this the case? Because verification takes place simply by wearing them. Wearables know you much better than other smart devices. The main function of a wearable device is to collect data from the individual user. After processing, this data can become a unique identifier for an individual— more secure than a fingerprint or passport.

This unique identity code could be derived from a set of specific and abstract data including heart rate, blood pressure, lipid levels, facial features, skin features, and even personal preferences. Using these methods, identity theft would become a thing of the past—and wearable data would become more secure. It would be possible to block access to data if the wearable device is not linked to your body, with significant possible applications in travel and tourism.

Consider how identification through biometrics could radically transform one of the most stressful stages of any holiday—the airport check-in. No matter how experienced a traveler you are, checking in for a flight can be a stressful and tiring process. You arrive two hours before your flight only to find a long queue has already formed at the check-in desk. You fumble with your documents, checking once more that everything is in order while you slowly shuffle forward. Your passport, boarding passes, and visas. For a moment you forget where you put your frequent flier card, but then you find it at the bottom of your airline-approved hand luggage. Someone ahead of you in the queue has forgotten their passport—people in the queue start to murmur in frustration at the wait. The hapless couple finally realize there is no way they are going to go on their dream vacation without that small document and trudge away, forlorn. A happy holiday ruined by a moment's forgetfulness. You shuffle forward, finally reaching the desk and handing over your documents. Then away goes your bag on the carousel—you hope you will see it again at the other end. You smile at the harried airline staff, grateful you got through the process without total disaster.

Wearables could transform the entire experience of checking in. Travel could become a seamless experience with no stress or missed flights due to losing a vital document. A simple wearable could link your boarding pass, biometric identity and other critical documents (e.g. travel insurance), reducing the risk of a misplaced passport or boarding pass. Wearables are far less likely to be misplaced compared to documents. Any wearable could act as proof of identity and provide access to an array of information about your trip.

The payment function of wearables saves the trouble of carrying multiple bank cards and membership cards, while providing additional layers of security. The wearable, linked to your body, could be used to open your hotel room door.

Timely warnings and reminders could be provided through your wearables while on the journey. From Big Data analysis in the back office, wearables would be able to give tailored information on traffic congestion, attractions, accommodations, and other points of interest, according to the habits and preferences of the wearer. Voice control will enable you to interact with your device. The wearable itself could become like an agent, building on what it knows about your preferences to help choose your rental car and deal with tickets or hotel bookings.

(2) Language barriers
International travel involves a lot of communication, some of which can be made difficult by language barriers. In some areas of the world a local guide may be needed. Juggling phrase books, tourists could struggle with the basics of ordering food or accommodations, particularly in remote areas. Language poses a barrier when accessing medical care or dealing with the police. Interpretation, when available, may prove to be beyond the means of the average tourist. The need for a device that can act as a universal translator is clear in this context, and some wearable devices are starting of offer a real-time translation function. Google Glass is one example which, though not yet perfect, can offer limited translation facilities.

Quest Visual developed an app for Google Glass called Word Lens, which allows viewed text to be translated into the user's primary language with the translation displayed using the Glass. For instance, during a holiday, a tourist might see a sign and ask Glass to translate it. This can improve safety and confidence for tourists in exploring new destinations.

The Google Glass version of Word Lens can be used in real time and has a dictionary of 10,000 words in each available language to support travelers.

The translation of text using online translation software is relatively simple, though the accuracy can vary depending on the language in question. Building this functionality into a wearable is relatively easy. However, the true potential lies in the use of wearables for simultaneous interpretation of conversations in the future. The need for this is clear—with over 6,000 languages on the planet, increased travel and trade, and the complexities of regional dialects, a smart interpretation system would offer assistance to tourists and business travelers alike. The potential exists for this kind of system to be able to learn through artificial intelligence. The potential also exists for users to improve the output by providing feedback that improve the algorithms. The more people use the system, the better the system will get. It could be based on an open platform and even be able to learn new languages, drawing on knowledge of the grammar, syntax, and vocabulary of languages in the database. The Universal Translator of Star Trek had some of this functionality—and if it becomes a reality it could transform travel entirely.

Removing language barriers would open opportunities for tourists to engage with different cultures. This is a practical application of how wearables can improve a tourist's experience. In the next sections we will examine how the tourist experience can be made more enjoyable.

2. Be your own travel agent and tour guide

Wearable devices offer great potential not only for data input but also for output. They could provide an almost endless range of possibilities to the tourism and travel industry. One of the potential impacts is a transition away from traditional travel agents and tour guides.

Traditional travel agents and tour guides may provide information, plan tours, and arrange accommodation and meals. They have a fairly central role in the traditional tourism model, particularly in certain countries where tourism may be less organized or for particular market segments who may prefer a more organized travel experience. How might wearables impact this part of the tourism sector? Let's consider what one wearable, Google Glass, may offer.

Google Glass can project virtual images and provide navigation based on voice recognition, essentially becoming a hands-free tour guide. With the potential for augmented reality, it could revolutionize the traditional tourism market.

Using Google Glass would liberate tourists, preventing them from getting lost or helping them access the local culture. It could give access to public transport timetables and allow for flexible traveling through internet-based hotel booking. With information about an attraction a literal blink of an eye away, the tourist will have access to a wealth of knowledge beyond that of a traditional tour guide.

Potential future opportunities may be presented by the development of tour packs, which would contain data about destinations, including local cuisine, a virtual tour, and other elements that may benefit the tourist. Available for download onto Google Glass, the pack would help the tourist to, say, briefly visit a destination such as Beijing's Forbidden City before making the decision to join the queue. Travel plans, possibly crowd-sourced from other users, may give interesting insights into options that different types of travelers may enjoy.

Google Glass could also translate local menus and restaurant reviews.

The tourist experience is a combination of comfort, food, accommodation, sightseeing, and entertainment. We can see the potential for new businesses that are dedicated to providing various types of data packs to serve a range of travelers' needs. Users only need to pay to download the relevant information, solving the traditionally time-consuming and costly job of travel planning.

Similar to the language packs previously mentioned, these travel data packs could also be updated at any time. For example, the tourist's choice of local food and restaurant ranking could add to the database to inform other users and help identify the optimal restaurant for the tourist's next night out. Crowd-sourced feedback might stimulate innovation among restaurants, targeting the different tastes of different tourist groups.

When such travel data packs are incorporated with the translation function, Google Glass could truly become a global travel agent and tour guide. No language or information barrier could stop you from enjoying a holiday anywhere on the planet.

The scope of wearable tourism is well beyond that of simple sightseeing trips. Being able to immerse oneself into the local culture and environment will allow a total escape from normal everyday life. Will we still need the traditional travel agent or tour guide? This is doubtful.

3. Travel light

I don't know about you, but for me packing luggage is one of my life's frustrations. No matter how well you think you have packed, it can still be a struggle to close the case. And I always end up forgetting something important, even if I've packed the proverbial kitchen sink. From clothing to cameras, the sheer bulk of stuff needed can tax even the best packers. However, wearable technology could transform what you need in your suitcase.

Consider a trip around East Asia, from the snowy mountains of Northern China to the sultry conditions of Singapore and Malaysia. Traditionally you would have had to pack clothes for the different climates—thick clothes for the cold areas and lightweight ones for the more humid tropics. Smart clothing offers a lot of potential for this kind of scenario. A piece of clothing could self-adjust to the local conditions, providing warmth in cold areas and cooling in hot areas. 4D printing technology offers the potential to create clothing that shifts shape under different environmental conditions, meaning one garment can suit a range of settings.

Capturing videos and pictures is part of the regular tourist experience—but how many of us have struggled through piles of vacation photographs from relatives or colleagues, unable to truly share the experience? Even with the best cameras you can miss the essence of the experience. Perhaps the emerging field of virtual reality recording may offer some potential here. This would involve the use of a device to record the experience and sharing that experience through immersive VR.

By simply putting on a VR headset, friends and family could experience the same sights and sounds. This may become the new way of recording future travel experiences, capturing the memorable moments of life to be relived again and again. You could use Oculus Rift, for example, to record your African holiday. Sharing that experience back home with your family and friends would be awesome, wouldn't it?

Maybe one day in future all we need to take will be a pair of smart sunglasses, a device that could be used as a camera, a video recorder, and a virtual reality headset. Recording and sharing fun holiday experiences could be ever so easy. It would preserve your travel experience with more fidelity than ever. It could completely recreate scenarios that are ready to be shared and potentially lived by others.

4. Virtual reality tourism

What on earth is "virtual reality tourism"? It sounds futuristic, but it is not so far away. Without leaving your home, you soon may be able to enjoy sightseeing in the world of virtual reality, like on the holodeck of the Starship Enterprise.

So far, the application of VR technology has largely been in the field of gaming. In the future, however, this technology will have a far wider impact on many industries including medicine, education and tourism. Tourism especially could be radically changed. The impact will be on both the marketing of tourism and the way consumers enjoy tourism.

The traditional marketing of tourist destinations is based on photos, videos, and other imagery to convince consumers that the destination is a fun place to be, and the choice to visit is one that the potential tourist will not regret. And yet most of these methods do not have very much impact. VR technology has the potential to disrupt the tourism market and radically transform the marketing methods used. Consumers would be able to have a genuine taste of the travel destinations through immersive VR experiences. Various VR-compatible videos could then be shot to promote certain destinations, targeting specific segments of the market.

Destination BC, a state-owned tour company for British Columbia, is one of the first companies tangibly using VR technology to promote tourism. They employed Oculus Rift technology to create a VR video called "The Wild Within VR Experience." The destination-marketing video was filmed using a custom rig—built with a 3D printer, no less. Seven specialized HD cameras were mounted around the rig, allowing footage to be filmed by helicopter, by boat, by drone, or on foot. The breathtaking views seen in the video were shot throughout the Great Bear Rainforest in the beautiful countryside of British Columbia, Canada.

At the end of 2014, Marriott International unveiled a new travel booth where hotel guests can explore the black sand beaches of Hawaii or the top of London's Tower 42 using the Oculus Rift.

Although sightseeing and touring could be one of the key applications of VR technology in future, it does not mean that people will no longer want to travel in person. But using this technology could make travel planning and scheduling much easier, taking advantage of VR preview and demonstration functions.

Marsha Walden, CEO of Destination BC explained: "We think virtual reality is a great fit for tourism marketing." The initial application for VR technology in tourism would be marketing, as it offers an immersive preview for users. But the other applications later on may include using VR to experience the entire journey.

For those who do not have the time or money to travel, or who are too ill or infirm, VR travel could meet their demand by using the previously mentioned

VR recordings. For example, maybe you have always wanted to visit Xinjiang, but have never found the time to visit. Virtual reality travel could make this possible. You could see the geographic features of Xinjiang, the vast farm and pasture lands. You could see what the local culture and food is like right in front of your eyes. And in the distant future you might even be able to taste and smell this food. The world would be at your fingertips.

It is particularly great for destinations some of us may never be able to reach or which are threatened by tourism congestion, like the ruins of Pompeii, the interior of the Pyramids of Giza, or many of the ancient Chinese landmarks. You could even visit the surface of Mars or the bottom of the sea. VR technology could make the impossible possible.

CHAPTER 15

EDUCATION

Things can be done much faster and more efficiently today thanks to mobile internet. Every aspect of our lives has been transformed significantly, including education. In recent years, we have seen many new education formats, online tutoring, online-to-offline education, personalized courses, improved self-learning and other advancements. All of these are trying to enhance, or address gaps in, the traditional education system. Such change is particularly needed in China, where educational resources are scarce and not equally distributed, and costs of education are high.

15.1 Online-to-Offline (O2O) Education

In China, O2O education has been attempted by many e-commerce giants including Baidu, Alibaba, Tencent, YY and Qihoo 360. They are using the capital market to test the O2O education waters, but progress so far has been disappointing. There are no clear business models for sustainable growth.

1. O2O education drying up

What is O2O education? To put it simply, it is a form of blended learning, combining online and offline educational resources. In terms of how to best combine these two types of material, each company has its own methodology. For example, the traditional offline education companies believe that for them implementing O2O education simply involves uploading their resources to the internet. They work on both content and the platform. However, they tend to reach a dead-end as the traditional players in the education market have little knowledge of how to exploit e-commerce possibilities. If the online aspect of their education system does not change the way they educate, it could backfire and be even less effective in terms of learning outcomes.

In comparison, internet companies can use their e-commerce and technology advantages to enter the O2O education market, but they tend to be too weak in the content they provide. NetEase Open Course claims that they are not seeking profit, but the reality is that they simply cannot find a way to commercialize their content. The content they provide is of very little relevance to the Chinese education system—and this reduces the potential revenue streams. YY, a voice-based software company, has also launched its own online education system. However, most of YY's user base are gamers and the conversion rate is rather low. Alibaba has also tried to tap into the online education market, launching Taobao Education. Since its official launch, the number of courses on offer has been less than a thousand.

The recent experience of O2O in China is shown in Table 15-1, which shows just how many failures there have been.

Table 15-1

Project name	Launch date	Functionality	Operations
72ren.com	December 2007	Online tutoring and learning website	Shut down
Coolban.com	April 2009	Connecting the online SNS community with offline study teams to provide effective English language learning schemes	Shut down
Hehu360.com	December 2010	Interactive online community for young parents to learn, exchange, record, and share information	Shut down
xue-du.com	August 2012	Online education sharing platform	Shut down
zhukaowang.net	December 2012	Online social education and e-commerce platform	Shut down
quer365.com	January 2013	Parenting and early education sharing website	Shut down
Cheers In	July 2013	Overseas education service customer-to-customer e-commerce platform	Shut down
Haniu International Early Education	October 2013	Parenting information, training and education service website targeting parents of children ages 0–18	Shut down

2. Problems facing O2O education

Online education relies heavily on self-discipline and self-motivation, areas where Chinese students are not so strong in general. Even the best international massive open online course (MOOC) providers only reach a 5% completion rate. Sina Education Channel Director Mei Jingsong pointed out that popularizing online education faces many challenges, including interactivity issues, Chinese habits of learning, incorrect assumption of all internet resources being free and more.

(1) Atmosphere

Educational attainment is particularly sensitive to the surrounding environment. People learn much better in a conducive atmosphere, without which the learning materials make little difference. If the majority of a class is keen to learn, those who are less keen could be motivated by the atmosphere and may gain a taste for learning as well. On the contrary, a good student could lose heart if placed in a class where nobody pays attention.

Existing online learning services still cannot provide a learning atmosphere that keeps people eager to learn and achieve results. Many pay the fee to start a course, but very few complete it. Learning can be tedious and exhausting for many, and it is hard to sit in front of a computer alone in a room to receive an education. If change is made to the learning model, it is very unlikely this business model will be a success.

(2) Motivation and pressure

Students in their last year before university entrance examinations have the highest stress levels among students in China. It is clear that stress increases as we grow older. And yet we need some pressure in order to achieve results. Traditional offline education is facilitated by tutors and teachers who supervise and oversee the learning of students. In China there are often penalties for failing to deliver assignments, missing lessons, or low exam scores. These penalties are weaker or non-existent in the case of online education. People are prone to take the easy road. Without supervision and punishment, laziness often gets in the way. Over 90% miss their online lessons and the completion rate can be as low as 4%.

At first sight, online education seems a good idea—allowing students to take responsibility for their own learning. However, the truth is that most students in online courses do not learn very much after paying the fee and are generally disappointed. The retention rates of such businesses are very low as a result.

(3) Lack of personalization

Traditional classroom teaching in China is generally delivered to classes of 30 to 50 students. Students all have a tutor assigned to them throughout their school life—from younger grades through university. Every student learns in a different way and has their own individual needs. A personalized study strategy may be particularly important in achieving learning targets for some students.

Personalized education is the future of education, and this may be difficult to achieve with offline learning. Online learning on the other hand could provide a better solution, although at the moment the online version is no better than offline. Many institutions only upload offline lessons to their websites for students to download. The true benefits of online materials are yet to be realized.

15.2 Wearable Education

Personalized and effective O2O education has only one place to go in future, and that is to join forces with wearable solutions.

The education system, which includes the format of lessons and approach to learning, will have to radically shift its focus. The objective will no longer be on the class as a whole, but on individual learners. Knowledge will no longer be about achieving certain quantifiable targets, but about the use of knowledge in new ways to reach higher levels of understanding about complex issues.

1. Immersive education

In the movie *Inception,* there are some interesting plot lines. There is a scene in the movie where Dom Cobb's wife was sitting on a windowsill, ready to jump in order to wake from her dream. (This is established in the story: a person in a dream will be woken by a fall.) Cobb explained repeatedly to her that they were in reality, not a dream, but she remained certain that it was a dream. She wanted Cobb to join her in the jump from the window ledge that would end the dream.

Why couldn't she tell the difference between reality and a dream? This is a profound question related to the human subconscious. Cobb planted the idea in his wife's subconscious that the world was a dream, although it wasn't. To deny something that you have personally experienced before is very difficult.

The same is true in learning.

We know some people have "book smarts" and others "street smarts." Individuals in the latter category often win out as knowledge from real-life experiences often registers better in the mind. This is what immersive education is all about. It uses VR devices to create an experience-based learning environment and structure.

For example, when a medical student is learning the basics of anatomy, a VR device could help the student visualize the circulation system to help the student take in the necessary information to improve as a medic. A VR device could also take a history student to the past to attend Lincoln's Gettysburg address instead of learning the speech on paper. Virtual reality technology allows information to run much deeper into the human mind.

Repeated immersive learning is the most promising direction for the future. It blurs the boundaries between false and real experiences. Simple knowledge memorization is camouflaged as personal experience that could register deep in a person's brain. In physics, the laws of motion could be learned in a much more interesting way. Students could be put in Isaac Newton's place to experience the process of discovery. The key to learning is understanding, a level higher than knowledge memorization. True understanding is the essence of learning.

What business opportunities could this educational approach bring? The advance of VR technology is the key one. Existing VR devices are not yet optimized for education purposes. The user experience needs further improvement. Ideally, we would like to see VR devices lighter and easier to wear, with many education and learning packs developed and ready to use. Current educational content is not adapted for use with immersive education. The development of various VR versions of content packs holds great potential in terms of businesses opportunities. Those able to develop innovative and easy-to-use resource packs will succeed in the future education market.

2. Downloadable learning
In *The Matrix* film trilogy, humans are connected into a matrix with tubes plugged into their brains. This matrix was controlled by a supercomputer ruling

over humanity. A small group of humans learned the truth and returned to the real world. Then they decided to fight back.

There is one scene in the movie where the resistance team needs to go back to the virtual world, but need more skills to defend themselves. The solution was to download this skill directly into the human brain—within seconds it is done and ready to use. In the popular Indian movie *PK*, the alien P.K. could learn any human language just by holding the hand of a person who speaks that language. These ideas seem totally fictional and unreal. But are they really that far off? Imagine the future where learning could be done by hacking the brain using a "plug-and-play" method. This would yield significant potential efficiencies.

This will only be possible through the use of wearable technology like a headset, for example. So far as I am aware, IBM has advanced quite far this direction already. The idea is to use a helmet-like headset with many sensors connected to the scalp. Using some form of neural currents (not fully explained), it accesses the human brain and transfers knowledge in a memorized format.

Knowledge packs, such as the entire history of ancient China, could be downloaded onto the wearable device and transmitted into the brain. How it might feel during the process of hacking your brain is hard to know. You might be amazed, when you find yourself like P.K., speaking a foreign language fluently as if you have always known it.

This learning method likewise presents great business opportunities in the development of resource packs. Both downloadable learning and immersive learning could co-exist in the future education market. The experience is different between the two. One is focused on immediate acquisition of knowledge, and the other is on adding fun to the process of learning.

CHAPTER 16

GAMING

Since ancient times, games have been an important part of human existence. Plato saw games as conscious mimicking developed by all youth (be it in the animal kingdom or among humans), to help them learn new skills and meet the needs of living. Aristotle took a more hedonistic view of games, seeing them as a type of rest and entertainment with no clear purpose, to be undertaken after the work of the day is over. More recently Raph Koster, the Chief Creative Officer of Sony Online Entertainment, defined them in the following way: "Games are the art of math. They are teaching us the system of themselves."[5]

Computer games are a digital extension of a tradition that extends back through the ages. Looking back, video games became popular in the 1970s, with the first commercially successful game *Pong* launched by Atari in 1972 in an arcade format. The first home gaming console was launched the same year, and by late 1970s home gaming had become popular. Gaming has gone through a number of changes, from the simple pleasures of shooting down aliens in *Space Invaders* to shooting a round of golf on the Wii to recently introduced VR gaming.

The future of gaming is clear: it will be an integral part of our lives, both at home and at work, in education and in our fitness routines. The current pressures on the economy and the significant development of social media are helping to drive the gaming industry forward into still-uncharted territory. In this chapter, we will examine how wearables are going to transform gaming.

16.1   Changes in Gaming with Wearables

After decades of evolution, gaming has become a second home to many. Game developers have seized the opportunities provided by emerging technologies to generate ground-breaking video games that transport people from their reality to a different world. Improvements in computing power, sound, and visual displays have all helped in this. Using joysticks, mice, keyboards, or touchscreens, we can interact with new virtual worlds from the platform games of Super Mario to the immersive experiences of Robo Recall on the Oculus Rift.

Web games are currently one of the most popular types of video games. First appearing in Germany, these browser-based games do not require users to download software and can be accessed from any location. They are popular among office workers, as they have a short loading time and are easy to close. There are also no computer memory or configuration requirements.

The gaming industry has been continually evolving and adapting. The main gaming platforms have shifted from TVs to Gameboys, PCs, and now smartphones. The gaming experience has improved through these different media. According to China's GPC (a gaming industry committee under the China Audio-video and Digital Publishing Association), mobile gaming reached over 330 million in China, with revenues of RMB 12.52 billion (USD 1.78 billion). This represented an increase of 394.9% compared to the same period in 2013.

For the same period, the population of app game users was around 130 million, a year-on-year growth rate of 3.7%, with a market share of 51.5%, reflecting a 17.2% decrease compared to the year before.

These statistics show us that the app gaming market has gradually been overtaken by mobile online games and other gaming formats. Since the Xbox One was launched in China, gaming consoles with strong connectivity and interactive features have appealed more to gamers. Offline games are no longer the market leader, making it even harder for less well-made games to retain users. In the coming years, there will be a stronger tendency towards "survival of the fittest" in the offline gaming market. Only the best games will remain. Therefore, they must look for new growth points for their future business, of which wearables are one potential area of development. This may transform the industry in ways never seen before.

1. Changing the user interface
The first change brought by wearables will be a change in gaming interfaces, with the replacement of keyboards and mice by the players themselves. Keyboards and mice have long been used as the major tools to connect users with the gaming device. However, no matter how well-made they are, these items still have a time lag between player action and the game system's reaction.

In fact, the best interface is no interface at all. Gamers perform best with their own bodies in reaction to the gaming situation. Wearable technology suits this perfectly as wearables are great at recording movement data. Gamers feel fully integrated into the gaming environment. Recorded movement data could be translated into gaming data, to be added to the gaming scenario. On the other hand, users could be playing games while walking and running. It may be possible to create non-gaming environment games with wearables.

2. Gaming formats
The formats of games are going to see changes too. Games will evolve from a virtual online world into augmented reality—a combined world of virtual and real-world elements. Online and offline activities will be merged into one.

On one hand, games are going to be part of our lives in all aspects through wearables. Traditional video games were, and are, largely separated from reality. Gamers have a real-life identity and a gaming persona. These two normally do not overlap. But wearables are going to change this. If you played a sports-related game recently, the smart shoes you wear would record your gaming movements and acknowledge that as exercise. And your daily jog could also be added to your online gaming account, giving you bonus points in your games. We will see the merger of online and offline activities. The smartwatches and smart bands we wear today have also incorporated this idea, but the benefits of gamification have not been totally realized. The manufacturers of wearables have gone some way towards the use of gaming social networks based on physical activity, but this has not been fully incorporated into other games.

On the other hand, the use of VR headsets in gaming provides an immersive experience which blurs the boundary between the real and virtual worlds. This is another way of seeing the two worlds come together.

In the future, games will be more experience-based. With the increase of different gaming formats and the growth in the entire supply chain of wearable technology, this could very likely be the case. In recent times, we have seen outdoor games, indoor interactive games, and family TV games increase in popularity. Nothing offers a better interactive experience than using wearable devices in gaming.

Gaming control is one of the key elements that determine the development of gaming formats. If this area is addressed well, perhaps television-based gaming could take off once again. To achieve this breakthrough, video game developers need to work together with wearable manufacturers. For game developers, it is important to involve wearable developers early in the game design process, in order to achieve the experience they want to provide. They may want to choose a specific gaming type and engage heavily in a form of "co-creation." It would be very exciting to see new gaming experiences on offer that meet the users' needs and exceed their expectations with fully supported accessories and gaming products. VR-enabled television-based video games, supported by wearable-based controllers, will provide entertainment in ways not currently imagined.

3. The games

The third change in the gaming industry will occur in the design of games themselves. Games of the future could be seen as a way to promote health and provide positive experiences rather than the overall negative image that gaming has today. The secret of such a change lies in the integration of the gaming industry with other industries. Gaming will become part of fitness routines, medical treatments, and many other aspects of life. The gamification of life means the games will be an integral part of daily life too.

The integration of game-based technology into health promotion will be greatly aided by the use of wearables. Gaming's movement away from the screen to augmented reality landscapes offers great potential benefits to players' health and well-being.

Scanning the horizon for future possibilities, we can see the potential for gaming to integrate with many other industries including fast-moving consumer goods, education, communication, IT, finance, and more. In the case of education, for example, it may be possible to use a mind-controlled headset device to combine learning with gaming, making a game out of a problem-solving lesson. Gaming could be used as an additional way of learning rather than a distraction. Gaming could be defined by healthy, fun experiences and improved satisfaction with life. Gaming could become integral to daily life—drawing on wearables to allow the game to go to places never before imagined. It may become a new "Game of Life." Everything you do could provide data to the game, allowing everything to be gamified and part of the fun.

16.2 Wearables as Gaming Peripherals

Among all wearables, VR devices are most favored by the gaming industry, already finding a warm reception in the gaming market. VR device makers also realize that the gaming industry, among all other industries, has relatively lower barriers to entry. Now let us look at the details of wearables currently used as gaming peripherals.

1. Oculus Rift headset
The Oculus Rift is a virtual reality headset developed by VR technology company Oculus, a company owned by Facebook. The display resolution is 1080x1200

per eye. It has a diagonal field of vision (FOV) of 110° and horizontal FOV of 90°. It can be connected to a PC or a gaming console through DVI, HDMI, or micro USB.

It was said that the prototype "Crescent Bay" also had sports and audio functions. The front of the headset was fitted with infrared LED sensors, with 8 LEDs on the back of the head, to allow for extended position tracking and increased tracking accuracy. The device supported a maximum of 90 frames per second.

Oculus has built an Oculus Home platform to help expand its range of online games. This platform supports all Oculus wearable devices and is compatible with Chrome as well as mainstream mobile devices operating iOS, Android, or Windows Mobile.

2. Google Glass

Google Glass is a head-mounted device providing augmented reality. However, most of the functions of Glass are not gaming related. Google Glass is more designed for assisting with daily life and work. Does this mean Google Glass is not useful in gaming? Not at all. The gaming potential with Google Glass is significant.

In 2013, Google launched GlassBattle, its first game app for Google Glass. GlassBattle was made by mobile app developer Brick Simple using the Google Glass Mirror API. When users played this game, they were able to view other users also on the game. Similar to the traditional board game Battleship, users deployed warships onto the game board and then attempted to destroy the warships of other players.

The major difference in the gaming experience provided by Google Glass is the fact that no physical controller is used. Users must use voice control for the game.

3. BrainLink headset

BrainLink is a thought-controlled headset device developed by Shenzhen Macrotellect Limited. It uses alpha and beta brainwaves to identify instructions, offers apps for users, and utilizes a Bluetooth connection. So far, BrainLink can be used for games such as Zen Garden, Mind Tower Defense, Mental Fruit Bomb, and others.

4. The Apple Watch
The Apple Watch is a smartwatch product which launched together with Apple's iPhone 6 and 6 Plus. Similar to most other smartwatches on the market, the Apple Watch has a flat square screen with a slightly chunky look. It offers voice communication, GPS, data transmission, and many other functions, but it has to be used together with an iPhone for full functionality.

Mobile gaming pioneer Gameloft soon released Apple Watch companion apps for four of its most popular games. Users are able to play from their smartwatch. Independent gaming developer Flying Tiger Entertainment also made an Apple Watch game called iArm Wrestle Champs, an arm wrestling simulation which makes use of the accelerometer in the watch to gauge arm wrestling ability. Admittedly smartwatch games are not perfect, but credit should be given to the gaming industry for quickly responding to opportunities with innovative solutions.

5. Zero smart footwear
Zero shoes are a wearable product designed specifically for the freestyle web game She Diao ZERO by the entertainment company Perfect World. Unlike smartwatches, smart bands or headsets, Zero shoes use the human foot as the sensor location. They work with one of the latest popular offline games by this company. This is a great example of exercise and simulation gaming. Zero enables data to flow between the games, and physical exercise undertaken by the player. Player interaction and exercise through gaming are the main features of this product.

Wearables are able to cover the human body from head to toe, and have infiltrated the gaming industry from desktop computer-based games to downloaded action games. Gaming obviously was one of the first commercialized areas for wearables, but much more needs to be done in terms of creating business models and improving device functionality. Many wearables remain communication tools for social networks or as health monitors. Applications within the gaming industry, however, are largely still at the conceptual stage, as the responsiveness and precision of control still pose many technical challenges to achieving a perfectly seamless integration of wearables into the gaming experience.

16.3   Virtual Reality and Gaming

In the movie *Her,* the lead character Theodore is a lonely man living in the future. Before he met and fell in love with Samantha (an AI system), he spent most of his evenings playing video games. In the scene, Theodore controlled a little character with his body. He was not using any VR or movement-based devices. It was an immersive 3D video game using projectors and glass wall displays. Although this is a science fiction scenario, it could be an ultimate version of what we call virtual reality technology.

As one of the most significant events in the video gaming industry, E3 has been the most high-profile gaming expo since 1995. It often sets the market benchmark for the industry. At E3 2014, 27 companies displayed their VR products. One year earlier, there had only been six such companies, mostly heavyweights in the business. For E3 2015, the keynote was based around the use of VR technology in gaming. Microsoft displayed its HoloLens VR version of Minecraft. Sony also announced more information about its VR display, Project Morpheus.

As with any field, revolutionary progress in technology leads to a qualitative leap in business. Gaming is no different. It is a business based on internet technology and the ultimate target is to offer gamers a perfect gaming experience. VR technology is seen to be the most promising innovation and gaming seen as a comparably easy market to enter, with the potential for new VR products to become commercially viable. The mesmerizing immersive experience of VR has driven both gaming giants and VR companies to face the challenges of merging the two. When Oculus announced their consumer edition of their VR headset, it as a big step forward in the commercialization of VR tech.

Microsoft has been the biggest winner in this race so far. It has invested in multiple areas of VR technology and has seized many opportunities. It launched its own VR device, HoloLens, demonstrated perfectly though Minecraft gameplay. Microsoft also announced its collaboration with Oculus, allowing gamers to use Oculus VR headsets to play Xbox games. It also partnered with Valve. Windows 10 will be the ideal platform for the Valve Vive VR headset.

Other industry players at the time were Sony's Morpheus, HTC, Valve, among others. They all worked to achieve the best immersive gaming experience with their devices. The Vive, for example, allowed users to walk freely in a limited space, providing an illusion of "reaching somewhere" physically. This function yields a temporary sensation of stepping outside of reality.

The true merging of VR and gaming requires much more effort in overcoming technical challenges. One of the major unresolved issues with existing VR devices is the nausea experienced after using these devices for a lengthy period. This is a major problem that dampens the user experience. Another issue is that consumer VR devices for gaming require much more affordable price tags. Higher prices would limit the user base to a much smaller group, as seen with Google Glass. If the industrial chain develops in the future to provide cheaper manufacturing costs, it would also drive the use of VR technology in a wider range of gaming activities.

16.4 Gaming Wearables Enable Ultimate Interaction

Mark Zuckerberg wrote about VR in this way: it's just like hanging out with your real friends anytime you want. He was quoted saying "Imagine ... studying in a classroom of students and teachers all over the world or consulting with a doctor face-to-face—just by putting on goggles in your home... By feeling truly present, you can share unbounded spaces and experiences with the people in your life. Imagine sharing not just moments with your friends online, but entire experiences and adventures."[6]

In traditional gaming, once the user leaves the screen, or puts down the handset, or exits the app, this pauses or ends the game. The same may not be true in the case of a game played on wearables, as the very means of interacting with devices are redefined. Anytime, anywhere, you can still have real-time interactions through a wearable device.

For example, Google Glass could be used as the scope for shooting games. You can have a fun game in the background of a real scene with Google Glass any time you like. Or you could discover virtual treasure hidden in physical locations. Wearable games can also be combined with exercise. In the 1990s Nintendo launched a Pikachu pedometer, which could turn real exercise and walking data into bonus points that could be exchanged for additional gaming content. Later this idea was further developed by Nintendo with the launch of Pokémon Go. With mobile and wearable devices benefitting from better sensors, this function has become more powerful and interesting.

In terms of gaming functions, smart bands and smartwatches obviously have more potential than Google Glass. Just imagine the monitoring functions that can be used to calculate your calorie burn during exercise, and the translation of that data into character elements within an MMO game. How cool would that be! There are of course much easier interactions with smart bands and smartwatches, such as the pedometer function. In fact, quite a few games have used this functionality in gaming already, including Walkr-Galaxy Adventure in Your Pocket. In-app purchases offer another revenue stream, though it is possible to use the app on a free basis. As long as you are willing to walk, you can redeem those steps for points in the game.

Other than these, there could be more potential gaming-related functions for smart bands and smartwatches. For example, using vibrations to give you a thrill during intense gaming, or alarms to remind you to collect your daily rewards. They could also be used as independent somatosensory devices for gaming controls. These are not science fiction at all. Many companies have already started working on such applications of this smart wearable technology.

Wearables could make gaming more diverse. Above all, they would make the gaming experience more realistic, to the extent that it becomes part of normal life instead of a separate activity. This could make a positive impact on life too.

This current age is one that has been dominated by smartphones. We have already seen how much we have changed because of that technology. In the coming age of wearables, there will be much to look forward to.

CHAPTER 17

FITNESS, THE FRONTIER FOR HUMAN-WEARABLE INTERFACE

F itness has never been more popular than in the past few decades. Yet, for many it is a struggle to get out of their seats and start exercising. Laziness or not valuing fitness enough can mean people don't exercise as much as they should—and this certainly can reduce the numbers that meet the recommended national guidelines for physical activity. For those who do put in the effort, there may still be barriers to getting the best outcomes—possibly because of a lack of knowledge about the best techniques for optimal results. Boredom, lack of support, and a lack of good coaching can all contribute to a loss of enthusiasm necessary for regular exercise.

It seems that it requires an awful lot to stick to a training regime. It is not a bad thing for business that consumers face significant challenges, as it means there are opportunities to exploit. But when there is no perceived need on the part of the consumer, what can a business do? Consumers may not be fully aware of their needs. These are what some might call latent, or potential, needs. Apple presents a classic example of spotting potential needs and developing products to address those needs.

It is similar to the fitness industry. Gyms are largely similar in their offerings and facilities. Apart from their physical locations, they are largely homogeneous. If you have a location with some equipment and a few fitness coaches, you can open a gym. And yet, fitness is still a subject of much discussion, which indicates some issues have not yet been fully addressed. The top reason behind such unmet needs comes from the fact that fitness and life are not well integrated. This raises the critical problem of how a business can help combine these two factors. Wearables certainly provide one solution.

17.1   Four Fitness Industry Trends Brought by Wearables

Many fitness wearables have evolved from simple fitness trackers into health monitors. The concept of using wearables to track health conditions is more accepted now. The next key issue after data tracking and recording is data mining.

Contradie's firm applies computational biology models to data collected from wearable devices to provide insights into what a user's body is doing and what it is trying to tell them.

Other companies—including Fitbit, Misfit (now part of Fossil) and Intel's Basis—are also doing similar things, while Google Fit, Microsoft Health and Apple Health services promise to provide universal storage for the collected personal health data and a deeper, more meaningful analysis. Wearables will no longer be simple trackers, but will provide solutions to a number of health issues.

1. Real-time health data tracking
The biggest difference between wearable-based training and traditional fitness training is the fact the wearables enable 24/7 tracking. The devices are generally lightweight, fashionable, and easy to carry. They monitor a user's heartrate, steps, sleep quality, body temperature, breathing, posture, and other body data. This data helps understand the heart's condition during exercise and recovery

time after exercise in order to plan more effective and suitable exercise routines, making fitness training more targeted and useful.

2. Supporting professional sports training

Wearables are particularly powerful in some areas of professional sports, particularly athlete training. They record crucial training data, which could be used to prevent injuries and generate better training results.

For example, there is a wearable device called the Motus sleeve, designed for those playing baseball or softball. It uses a lightweight sensor inside a compression sleeve to collect biomechanical data with every throw. The athlete can use this device to correct his or her technique, and this may help avoid injury to the ulnar collateral ligament (UCL). Combining accelerometer data with a gyroscope allows tracking of how the throwing arm moves, all of which can be downloaded to a smartphone and uses Bluetooth. Based on the arm movements and basic biomechanics, the Motus App is able to calculate the elbow torque on the UCL. It can track dozens of pitching and batting metrics, including hand speed, workload, power generated through the hips during a swing, and elbow torque.

This type of wearable is much more refined in terms of tracking and can command higher prices as a result. They are only used by professional athletes. Wearable makers may be able to produce these products for ordinary consumers who are amateur players of sports like baseball, football, basketball, or rugby. With higher performance and lower price tags, there is potential to position these products to different markets in the future.

3. Emotion trackers

As wearables are closely attached to the bodies of users, they are the smart devices that have the best knowledge of users through precise data monitoring. We know that emotions can affect the physical condition of an individual.

Made by Sensoree, the Mood Sweater is another product for emotion tracking. People can know your mood without asking. The sensors on the product can transmit emotion data to the collar of Mood Sweater, with the collar changing color according to whatever mood is detected.

British Airways also conducted testing on a Happiness Blanket. The blanket could monitor the user's mood through neural sensors on Bluetooth connected earphones. The blanket emits blue light to indicate happiness on the part of the user and red light when the user is unhappy.

There are many similar products. I predict that these products could be most useful in reminding users to keep a check on their stress levels, so that people may take precautions and adjustments before more significant physical or mental health issues arise.

It is rather important to control stress for heart health and to maintain a healthy body fat level. Maximizing one's mental stability is a key to remaining in good overall physical health.

4. *Burning calories*

Most people into fitness will understand the concept of calorie counting. It is very useful to keep track of the right amount of exercise to achieve good fitness levels. For those who do not have the habit of regular exercise, but want to develop such a habit, wearables could help improve self-discipline. Such users are keen to know how many calories are burned as the result of exercise. Most wearable devices sold today are not accurate in the way they estimate energy use, therefore the data results are not necessarily reliable. Existing devices currently use movement acceleration to determine the exercise type, then combine this with age, gender, body weight, and other data to calculate calories burned. This is obviously not an accurate method, nor does it help to calculate calorie intake from food.

Whether wearables will continue to be used by consumers in the future is determined by multiple factors—the benefits of around-the-clock tracking, the comfort involved in wearing the device, user friendliness, data accuracy, and more. Users need to be assured that the wearable can help them live healthier. Apart from well-developed activity trackers, more wearable formats will enter the fitness market. Business opportunities will be available for every part of the human body. We will see smart shoes, smart socks, smart knee pads, and smart wrist pads. There are promising marketing prospects for smart sports clothing, which could be worn by many.

17.2 Smart Clothes Used in Gyms

Wearables are no longer novelties at the gym. People can be seen wearing chest straps, smartwatches, or smart bands to track their performance and hit particular targets. However, there are problems with existing products that are hard to ignore. One issue is awkwardness. For those who have never worn watches or bands, having an additional item on the wrist may not prove comfortable. To address that need, more smart garments have begun to appear on the market. Smart tops and tracksuits are as easy to wear as normal clothes, but they are also able to track heart rate, breathing, and other biometrics associated with physical activity.

Smart clothing is predicted by some to become one of the best-selling wearable technology items. The rapid increase in sales—from as little as 10,000 units shipped in 2013 to 100,000 shipped in 2014 and up to 10.1 million in 2015—shows the potential for this segment of the market.

OMsignal and Hexoskin are two major players in smart clothing. They are both based in Montreal. The Hexoskin smart shirt has been adopted by Olympians and NASA. OMsignal developed a new smart shirt series with Ralph Lauren for the US Open tennis tournament. In 2015, some new companies like Athos also entered the arena.

These smart clothes share one feature in common: there are concealed sensors implanted inside the garments. Smart clothing is not like other wearables, as the materials feel normal—there is no need for additional items to be worn that could get in the way. Using the technology without realizing it's there is in some ways the ultimate goal of the wearable technology industry.

2014 saw wearables thrive throughout the year. Growth can be particularly anticipated in smarter fitness activity trackers. The collection and analysis of tracker-based data will change dramatically as a result.

At the Consumer Electronics Show (CES) in 2015, wearables were the centerpiece of the show. From smart garments to smart bandages, these products all attracted a lot of interest. The collection of biometrics data, including breathing rates and metabolism, is increasingly comprehensive and professional.

159

17.3 Fitness Training for All: Exercising While Playing

Bing Gordon, a partner at KPCB, the largest venture capital in the US, once said that every startup CEO should understand gamification.[7] No matter a startup's area of business, awareness of gamification during strategic planning is very important, especially when gaming is the new norm.

In healthcare, where medical treatment and routine care can prove tedious and monotonous, involving gamification can encourage users to participate more willingly. As a result, they can form healthier lifestyle habits and better utilize the health functions of the wearables.

The fact that gamification rewards people for a desired behavior is pure human psychology. Combining this element with medical treatment means patients are more likely to act willingly and enthusiastically to improve the management of their health conditions. The gamification of healthcare has incredible potential to benefit patients, businesses and the wider society.

1. Games release the desire to compete for fun
Once this desire is combined with wearable medical devices and their apps, it could form a community with high rates of retention. For example, when step or calorie data recorded by an activity tracker is shared through social media, it makes the data viewable by others. Once such information is shared, a sense of competition could start to appear among a circle of friends. An individual may want to take another short stroll if someone they know is slightly higher in their step count ranking.

As AI develops further in future and wearables are able to read you better, they may even talk to you when they know you're receptive to messages like "That person has gone ahead of you." Then you might stand up and figure out a plan to regain your place. Such competition would make daily exercise routines much easier to stick to.

2. Gamifying non-communicable disease management lets patients self-manage diseases in their daily lives

Ayogo incorporates games into Healthseeker, an app specially designed for diabetic patients and children prone to developing diabetes. Users may start by choosing a life target to complete, then win badges and medals by completing various tasks.

Any habit requires time to form, particularly habits for a healthy life, often requiring additional drive to deliberately repeat certain activities. Nannying, dictating orders, soft reminders, or punishments will not necessarily be effective. Gamification is different, however. Users form habits unconsciously through the playing experience. It is not only enabling but also entertaining.

3. Gamified health management could help collect better data for medical research

The role of social media in spreading popular games is significant. This could be used by researchers as a way of helping to study and track health management data from diabetic patients, for example. Researchers could look for communities of health management games especially developed for diabetic patients. Data collected from these communities would have a larger coverage and be more complete. Apart from biometric data, there is a potential to study data from social media on the content of messages and analyze the role of the health service. It is possible to analyze the sentiment of messages on social media on certain topics, for example.

The clustering of patients in social media would not only be good for medical and pharmaceutical researchers for their targeted studies, it could also be useful for patients to learn from each other and to obtain peer-to-peer support. It may also be possible to market online consultation for patients with similar medical conditions, using social media channels.

Of course, the main challenge facing gamification of health management is user stickiness. Although games could release the user's internal drive to continuously monitor their health and make necessary adjustments in life, the big question is how to retain users and make them loyal fans of the games they play. This can be tackled as follows:

(1) Upgrading

No good games remain consistently attractive without regular upgrades. Fast reiteration is the key to retaining user attention these days. Health management games need to learn this lesson, in order to effectively intervene in users' daily lives.

Apart from making the games interesting in themselves, useful in-game databases and links to other resources are also important. These could draw on medical databases, tips from social networks, existing guides on health management methods, and so on. Users always value scientific, up-to-date, and effective health information.

(2) The social element

Incorporating social elements in games is common today. Most users prefer to play games with others. Competition also makes games more fun. Users also like to share experiences. Therefore the social dimension of gamification is crucial and highly valuable.

As previously mentioned, Ayogo developed the software Healthseeker for diabetic patients and children at risk. As it is linked to Facebook, that platform provides a massive potential user base which could easily bond together as a supportive and competitive social network.

Social games also increase the sense of achievement during competition. While playing games, our bodies release dopamine, a chemical that stimulates the brain and makes you feel happy. Such feelings could then encourage users to carry on, forming a virtuous cycle and improved health.

(3) Strong and powerful incentives

When we talk about strong and powerful incentives, we talk about, for example, points collected in the game that can be redeemed for real-world prizes or cash. Mango Health is one of the schemes under which users can be rewarded with real money when they achieve certain targets. But health management games are different in some ways from other types of games. The aim of traditional games is fun, and rewards are directly usable in the game. But health management games are trying to persuade users that it is worth employing this management tool for better health, encourage user retention, and allow for the collection of a sizable volume of data. Developers could then use the data to assist further commercialization.

UnitedHealthcare is a company out of Minnesota, in the US. Their "Baby Blocks" program is an interactive incentive program developed using smartphone technology to encourage regular visits to the doctor during pregnancy. There are over 50,000 pregnant women from over seven states participating in the program. These moms-to-be can unlock new game levels by making antenatal appointments. After completing some key appointments, they could receive gifts including maternity clothes, baby accessories, and gift cards. According to UnitedHealthcare, there were 2,296 active users of this software in 2012, with 7,098 antenatal checkups. An average of 3.1 levels were unlocked per person.

Other incentives include lower insurance premiums for those who exercise regularly, demonstrate healthier lifestyles, or show improved medical conditions. Free online and offline consultation for users of a certain level is another incentive. Well-designed game levels and incentives could be the key to attracting and retaining users. Physical incentives are particularly important to drive users to unlock more levels by completing more tasks.

(4) Privacy protection

Data safety and privacy protection have become increasingly critical in the age of mobile internet. The gamification of health management faces the same challenges. All involved parties—software and hardware developers, insurance companies, hospitals, and other medical service providers—could have access to users' private information. A patient with a non-communicable disease may be a willing participant in a gamified management scheme, but still not want their personal information or even their condition to be shared. Certain diseases, including hepatitis or HIV/AIDS, have significant stigmas associated with them. Without careful design, the social interactive side of gamification could compromise the safety of personal information.

Apart from the risk of potential data leaks, participants need to work together with businesses to develop a user-centered data sharing agreement.

For example, if we want to develop a gamified health management app for diabetic patients, we want the app to monitor blood sugar every day to figure out the reasons behind elevated levels, so a healthier diet and exercise routine could be suggested. Users could also get points for completing these tasks. When the tasks are completed, they'd be presented with the choice of whether to share this with friends on social media, and should be able to choose an option based on their privacy preferences.

Whether a device or software can flexibly and effectively protect privacy will be one of the core criteria in evaluating user experience in the future.

CHAPTER 18

ADVERTISEMENT

L ike it or not, advertisements are everywhere. We can hardly avoid being influenced by them as we shop, so businesses are always looking for better ways of encouraging us to buy their goods and services.

What about wearable technologies? Have they been ignored because of the limits on screen size? By no means. Advertising agencies have been studying this opportunity for a long time. The key word in advertising is precision. Wearables provide the most precise information about the user, because they monitor the user around the clock in a more detailed way than any police stakeout could watch a criminal. Such information allows companies to develop more personalized real-time advertising with precise delivery.

18.1 Who Are the Movers and Shakers in Advertising with Wearables?

1. Wearable advertising engine

Tecsol, an Indian software company, introduced its advertising solution for wearables in 2014. They used the Motorola Moto360 smartwatch to display advertisements. Possible uses include letting the user know about a nearby coffee shop while walking on a street or, unprompted, telling the user the weather forecast as they head for an appointment on their calendar.[8]

Tecsol built a basic MVC model of a Cloud-hosted ad engine. A basic static advertising image can be uploaded and pushed to the wearable device. This pops up, and the user can click on it or dismiss it. The watch then sends data back for analysis.

2. Virtual mock-ups of ads on wearables

"Any device with a screen allows for an interesting opportunity,"[9] explains Atul Satija, vice president and head of revenue and operations at InMobi, a maker of mobile ad tools.

InMobi has a team of developers creating virtual mock-ups of ads on smartwatches, head-mounted displays and other gadgets to get a feel for how they can serve as a platform for advertising agencies.

Millennial Media and Kiip have also joined the search for viable wearable-ad technology, underscoring the appeal of the devices as marketing platforms.

18.2 Advertising with Wearables

The business opportunities brought by wearables are based on the valuable, unique data they collect. After processing and analysis, more useful and detailed user information could enable marketers and advertising companies to push more precise ads to the consumer in a new and exciting way. This would take advertising to a whole different level in terms of targeting and personalization. This is significant for the future of mobile marketing.

1. Smart glasses
"Knowing where I am is interesting. Knowing what I'm looking at or studying for 3 to 4 minutes is more interesting."
— Julie Ash, analyst at Forrester Research Inc.[10]

The use of eye tracking in smart glasses offers great potential for understanding the shopping habits of users—detecting what users are looking at, what catches their eye, and what they think about when they are shopping. The advantages of this advance in wearable tech is clear.

Google Glass has a patent for a gaze-tracking system, which follows a user's gaze to know the user's feelings. It can also generate a gazing log, tracking the identified items viewed by the user and the user's emotions at that moment in time.

Compared to other smart glasses Google Glass is apparently most competitive in the advertising business, as it has the best user data support. Google Glass can provide information on nearby restaurants based on user preferences, or tell the user where their friends are eating. Coupons or discounts can be offered based on Big Data analysis.

Google Glass also has a better human-machine interaction experience than most others. This makes Glass users more likely to accept ads on the device. There might be voice-controlled interactive ads and user-selected ads with hands-free wearables like Google Glass.

Smart glasses and smartwatches have the largest screens among all wearables, but the largest are still relatively small. If advertising companies try to fill the screen area with ads, they will not necessarily be well-received. Although Google has obtained a patent for placing ads on Glass, they have not yet acted on it. They even claimed that they will not consider advertising on Google Glass. Google seems to be very cautious in advertising on wearables, but they are also likely to be the first to try this approach. They may be an early adopter in wearable marketing just as they were with wearable technology and smart home devices.

2. Smartwatches

No genuine smartwatch ads have hit the small screen yet, but many marketers are attracted to this idea. The Apple Watch has been visualized as a billboard on the wrist by many mobile advertisers. Limited by the size of the screen, ads on smartwatches will be very different from those on smartphones.

A smartwatch may look like just another screen to conquer after television, computers, and smartphones. But it is a new frontier between consumers and advertisers.

John Havens wrote in *Hacking H(app)iness,* "It [your smartwatch] might say, 'Your pulse just went up, lay off the coffee.'" Havens also foresees a slightly more insidious use for smartwatches. Retailers, he speculated, could monitor your pulse as you walk through a store. If you see an item that makes it race, they could then present you with an offer.

Such an advertisement delivery format would redefine "preciseness" in advertising. Traditional precise advertising is based on Big Data analysis. By tracking your internet search history, webpages pop up related ads. The advertiser may not truly know what you like, or whether you bought the products or not. The mind-reading ability of smartwatches, however, is in a different league. They could be used to measure user preferences through heart rate monitoring. Precise preference data could be generated through analysis of the accumulated data of multiple users over time.

Wearables are already collecting a great deal of biometrics. Through pulse tracking, smart wrist-mounted wearables have great potential in healthcare, but will they be fantastic for advertisers keen to understand consumers' feelings? It could be a great application of this technology, but in terms of whether it will be as useful in advertising as in healthcare, we will have to wait and see.

3. Mind reading at your fingertips

Wearable device maker Personal Neuro Devices has a tagline: "Your mind at your fingertips."[11] What does this mean? It indicates that future ads could be pushed to users as a result of reading the minds of users. When you feel low, wearables may push ads for chocolates or music albums you'd like. When you feel hungry, the device could read your brain and decide which cuisine to recommend—Chinese, Italian, or junk food for example.

The potential for mind reading via technology has been exhibited in a number of different studies in recent years. Functional magnetic resonance imaging technology has been used by researchers at Cornell University to help decode images in the mind. It is possible to work out what people are thinking about based on these scans of the brain. Similar studies in Japan have shown promise in helping people with locked-in syndrome communicate.[12]

I believe this must be thrilling news for advertisers, but for users this may raise concerns. If brain scanning and mind reading technology advances to the stage where private thoughts are immediately known, this would be an ethical and moral minefield. This new concept of marketing is called neuromarketing.[13] It is indeed a challenging concept.

Wearable marketing may take on many forms in the future, but so far it seems more practical to focus on GPS-enabled advertising of local points of interest, with potential for targeted discounts and integrated payment through wearables. This is the easiest marketing approach for users. After all, nobody would be offended by a discount offer, would they?

18.3   Challenges Faced by Wearable Ads

1. Advertising media
Google predicts that future ads will be placed in more unexpected locations, on radiators and refrigerators in the home, on car dashboards, through glasses and on watches. We can imagine using fridges or dashboards as advertising media, as they have relatively more space. However, wearables are not quite the same as these surfaces.

First of all, existing wearable devices all have very small screens, or none at all in the case of smart bands, smart rings or smart clothes. How can these be used for advertising?

An American startup released a smartwatch that is able to project information onto the back of the user's hand (Figure 18-1). With a built-in micro projector,

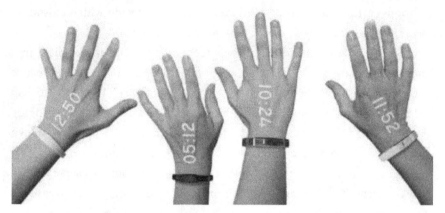

Figure 18-1 Smartwatches that project information onto the back of hands

it can display the time and other notices from the linked smartphone onto the back of the hand. For small screen or non-screen wearables, image projection could be a solution that provides space for revenue-raising advertisements.

However, this still doesn't solve the entire issue. The overall trend of future wearables is to be less visible, with the sensors more closely incorporated into human bodies. When this happens, there will be nowhere left for information projection. Where can the ads go?

The use of voice is one answer. The next stage of human-machine interaction will be voice control. Users will probably shift from the receiving side to the requesting side for ads as well. For example: you want to buy new clothing, so the hidden device detects your desire and scans available products in the background based on climate, temperature, user body shape, user preference, and price range. The filtered search results would then be pushed to the user to browse. How could this be displayed? Projecting such information in a 3D space with virtual reality technology is what we can expect in the future.

After voicing the request, "I'd like to buy some clothes," your own virtual image will soon be trying on different clothes in front of your eyes. Instead of viewing the clothing on models, seeing them on your virtual self could make your shopping decisions easier and returns less likely.

Therefore, the ultimate display option for wearables might be based on virtual reality technology, which is flexible and not restricted by any physical screens. We can view physical screens and projections as necessary steps in the march to VR, but these steps could take quite a long time, as core VR technology is a tough battlefield to conquer.

2. Consumer attitudes toward advertising
The modern concept of the advertisement was born in the late 1970s. Since then, the format, delivery, and media of advertising have all changed dramatically. Ads today find every possible way to present themselves before consumers, and yet the overall consumer attitude towards advertising has been going in the opposite direction.

OMD published a consumer report recently on attitudes towards advertising. In this report, it points out that consumers have extremely diversified attitudes towards mobile advertising. The majority (89%) have negative feelings towards mobile ads, and yet 75% also think the ads are fun. As many as 94% even say they believe ads are necessary.

Clearly people have mixed feelings about advertising. In general, advertising is accepted, but we don't want to be bombarded with advertisements. So far we don't have clear statistics on wearable ads, nor do we know what ad format users are more likely to accept with these devices. One thing for sure is that compared to traditional screens, wearables are more personal. Users are more likely to consider wearable ads more intrusive than ads on any other media.

As personalized and precise marketing has become the new normal for advertising agencies, some buffers remain between users and advertisers. Yet users still feel that their personal space has been invaded. Wearables bring a new relationship between consumers and advertisers. And the devices will be responsible for displaying ads, setting ad preferences, and making ads more like functions that support a better quality of life for users and reflect their own values.

IDC's recent study shows that consumers are most welcoming toward products recommended through social media and their circle of friends. In other words, as long as the products are truly desirable, it doesn't matter so much if it reaches them in ad form or not.

Wearables are likely to further blur the boundary between marketing and life. The relationship between consumers and advertisers will also be redefined. Maybe one day, even the word "ad" will no longer be used in the traditional sense.

3. *The inevitable conflict between Big Data and privacy*

The objectives of commerce and privacy are often in conflict, especially so in today's Big Data world. As data processing power continues to improve, whoever wants to make use of the collected data can figure out rather comprehensive personal profiles for any given user. In a way, everyone is being scrutinized by thousands of eyes, from all different angles, around the clock.

Users, on the other hand, enjoy all the convenience and personalized services made possible by Big Data, but worry about violations of privacy at the same time. This anxiety was clear in the case of Google Glass. Even though the device was not guilty of anything, people were still worried.

People are sensitive to the risk of their privacy becoming compromised with the mobile internet, and such fears only become worse in the case of wearable technology, as the core of the technology is to generate value from personal data. Wearables also present a different marketing platform. Advertising would be more invasive, making privacy issues an even bigger problem.

Capitalizing on Big Data means battles between users and businesses are going to be inevitable. Businesses want to seize the opportunities made possible by knowing more about their customers. Users are going to fight for both their privacy and the Big Data-enabled benefits they enjoy. In fact, this is an unsolvable conflict. The balance will constantly shift between the two sides. Advertisers need to know the true thoughts of consumers in order to deliver precise ads. When the true thoughts cannot be openly requested, all sorts of "other means" come into play. Consumers' fears, on the other hand, largely stem from the unknown trajectory of this technology. No one can tell how far businesses are going to push the boundaries. And this fear makes the privacy conflict a potentially drawn out affair.

Finding the balance between Big Data capitalization and user privacy protection is one of the key issues that needs to be addressed in the age of wearable technology. The "right to be forgotten," under EU law, allows users to delete information that could potentially violate their privacy. This is the first step the EU took to protect the privacy of the general public. It may not have significant impacts, but it is a reminder to all. The commercialization of Big Data is unstoppable. Privacy protection is also receiving more and more attention. In the future, individuals may be empowered by law to protect their privacy and enjoy additional rights.

With the radical change in format, value, and media of advertising in the age of wearables, the sky is the limit for potential value across the entire wearable industry.

CHAPTER 19

SMART HOMES

Rapid advances are being made in making our homes "smart" —from smart lighting solutions to smart refrigerators, smart entertainment systems to smart cooking devices, smart heating systems to smart key systems, there is hardly a room in the house not subject to the "smart" revolution. The interconnection of devices, using home hubs, is giving rise to new opportunities for business development. The era of the Internet of Things is upon us.

Leading brands such as Samsung and Haier, with their U+ Smart Home Platform, have been particularly active in developing smart products for the kitchen. Smart fridges offer a range of functions—cameras allowing inspection of goods inside when you are at the supermarket, to smart ordering systems, to integrated systems that display recipes on the surface and automatically set the oven to the desired temperature.

One example is the Haier Aqua fridge from Japan. This appliance is equipped with 32-inch high definition LED screens on both doors, which can showcase the freshness of the food stored inside or simply operate as TVs. It is WiFi-connected, for easier online grocery shopping. There are other apps and services available through Android OS. This smart fridge takes the simple refrigerator to a whole new level. The additional functions and user experience can only be described as revolutionary.

This attempt by Haier confirms that the shift to smart homes is becoming mainstream. The fridge represents low-hanging fruit, in terms of using doors as displays, and there is great potential for using these spaces in innovative ways.

19.1   Man, the Center of the Internet of Things

Laziness is part of human nature. We naturally prefer enjoyment and activities that require less effort. That preference has also driven the development of new technology. When we examine the development of the mobile internet, everything was anthropocentric—humans were at the center. Looking ahead to the Internet of Things, it is even more so, making life better and easier. The first step is to build smarter homes that can interact with wearables.

Wearables shorten the distance between humans and devices, enabling control through novel interfaces in the time of AI. Mind control is moving from science fiction to reality, with the integration of digital and biological technology and the use of Big Data and cloud computing. Based on the brain control information database built through neuroimaging, brain implants could be placed into corresponding control locations to intercept or record (or stimulate) signals transmitted from brain neurons, in order to acquire specific information or execute orders given by the person with the brain implants. Using this technology, implants could be placed at different locations in the brain to achieve different purposes.

Man could then become the center of an AI terminal. The smart home would not need a computer as a control center, as the human brain would be the integral part of the system. Your thoughts would initiate instructions to control the appliances around you.

19.1.1 A 24/7 AI world

Science fiction films often present dramatic pictures, with a super brain in the future surrounded by AI screens everywhere.

Low vibration, smart alarm clocks might be able to set themselves half an hour before your ideal waking time, in combination with sleep tracking data, to ensure a good start to the day. Breakfast could be prepared totally by machines. Well-balanced, nutritious menu options could be displayed on a virtual reality smart screen.

Before you head off to work, your smart device could plan your commute based on real-time traffic flows, drawing on Big Data and cloud computing. An optimal route would be planned with a nearby parking space pre-booked, to minimize waiting time.

Once you leave, the smart home could automatically enter an empty mode, with lighting and air conditioning switching to energy-saving settings. Some appliances could start preparatory work or take stock, so food supplies are ordered online for timely delivery. A cleaning robot could start its work around the home, and switch to stand-by mode once cleaning is completed. Before you finish work, you indicate your meal preference through a remote device to enable AI appliances at home to start preparing dinner.

When you arrive home, a smart device could open the garage door for you and a biometric ID recognition security system will let you in. All ambient settings will be adjusted according to your preferences, including lighting, sound, AC, TV or wall displays, creating the most comfortable home environment for you. If you fancy a party, there is no need to go out. All your smart home appliances will prepare a fantastic party night with delightful food and drinks as well as a suitable party setting. After a nice soak in the bath, you could greet your friends and family using VR video briefly and go to sleep. Your smart home could again adjust your environment to best ensure quality rest, with on-going sleep monitoring throughout the night. Smart homes could provide emotional help as well as practical convenience. An interactive AI voice system could be your assistant and friend to help solve issues and provide suggestions.

The smart home and the Internet and Things will connect all vehicles, appliances, and devices. They will all be equipped with sensors, processing information and responding to situations. There will be direct conversation between devices and people.

19.2 Wearables: The Best Smart Home Terminals

At the moment, most smart homes are controlled through smartphones or tablet computers. The smart home business is in its infancy, and smart homes of the future could be connected through wearables. Real-time data from the human body could be central to the efficient operation of smart homes.

Haier, a pioneer in smart home appliances noted above, recently developed a smartwatch that can control air conditioning units, one of the first of its kind in the world. Users only need to issue simple voice commands to control the air conditioning through this watch, including switching the unit on and off, adjusting strength, controlling the temperature, and more. Compared to the more common smartphone app controls, this is easier to use. It is a step forward from the general "device+app" model to smart durable goods.

The most revolutionary change in the integration of wearables with smart homes is the move towards smart appliances becoming less visible in daily life. They will still be physically present, but human-device interaction will be increasingly driven by processed data derived from smart wearables. Imagine, for example, going for a jog and returning home. Instead of needing to open the door, switch on the AC, and waiting for the temperature to drop, all these actions will no longer be necessary thanks to smart devices. The smart door would recognize you and open, while the ambient environment of the home would be automatically adjusted according to your body data before you arrive. Hot water will also be ready to use in the bathroom. Ultimately, there would be little or no conscious interaction needed.

Both smart home appliances and wearables are smart technology, therefore it is inevitable that these two will eventually merge. As an extension of our bodies and intelligence, wearables embody the interaction between humans and things. In the context of smart homes, wearables could be the key to opening up the wonders of smart appliances. They significantly shorten the distance between users and devices, enabling "zero interaction" in some cases. We can expect to see existing smartphone- or tablet-based smart home appliances being totally replaced by wearable-controlled devices very soon.

19.3   The Advantage of Wearables for Smart Homes

Currently smart homes are still at an early stage of development and require people to directly instruct the device to operate in a particular manner. The simplification and portability of controlling terminals for smart homes are very important. So far, most smart homes are controlled by smartphones and tablets. If the controller is converted to wearables, the user experience will be completely transformed. I believe the strengths of wearables lie in following aspects:

1. Simpler operability and improved user-friendliness
Compared to smartphone-controlled smart home systems, the use of wearables will make smart homes easier to run. Measurements from the human body or the natural movements of the body (blinking, waving, and so on) could give control signals that trigger actions. This is much easier than using fingers to press buttons or flick through menus, and would significantly reduce the time spent interacting with devices.

2. A 24/7 attachment to the body, with no limits on time or space
In the same way that smartphones are much more portable than PCs, wearables are even better for carrying around. You can be glued to the smartphone, but you can't go to bed holding it all night long. Most wearables, however, can be worn around the clock. The value here lies in potential applications for continuous home care for patients—linking to smart home devices can improve your quality of life.

3. Computing abilities make smart life more personalized

The increase in computing power we have seen, and will continue to see, will make the application of biometric data to different needs ever smarter— recognizing patterns will allow for better interaction with the Internet of Things. When such data is incorporated into the design of the smart home, it will fundamentally improve the user's experience. It would be personalized to fit the needs of the people living there exactly.

4. Full human-device interaction

Currently smart home devices connect to each other, but do not so readily connect to the human user. All technical discussions on computer bus, WiFi or radio-frequency technology are all based on connections between hardware devices. But the core function of a smart home is allowing smart products to serve the users. The effective link between humans and products determines the level of connected and controlled automation. Human biometric data could serve this purpose, both on the move or in sleep mode.

CHAPTER 20

PUBLIC ADMINISTRATION

The possibilities for the use of smart wearables in public event management are significant. There was a case of a woman living in Calgary, Canada, who used her Fitbit data in a personal injury case to demonstrate that her activity level dropped after an accident. It is important to note that the data had been processed and analyzed by an independent third party Vivametrica before submission to the court. This was the very first time in legal history that data from a personal fitness tracker was used in court as evidence.

This case allows us to foresee even more applications in time, from integrating personal data from wearables, especially in the public administration, e.g. crime management, criminal investigations, city planning, and surveying public opinion. Wearable technology will become increasingly useful and result in cost savings for various government agencies.

Professor Alex Pentland, a pioneer of wearable technology cited by Forbes as one of the world's top seven experts on Big Data, explained some of the underlying opportunities. He said that wearable technology could play an immeasurable role in many areas including health, finance, urban planning, crime prediction, and more, providing an image of a future driven by data.

20.1   Identity Verification: The Killer App of Wearable Devices

We have seen more and more methods of identity verification, along with increasingly complicated safety measures. Many smart devices are able to quickly and safely extract human biometric data like fingerprints, heart rates, facial features, and others to verify the user's identity. Compared to traditional encryption methods, encrypted biometrics are far superior and will gradually take over as the mainstream identity verification method for social media, smart devices, and some payments.

To make this method absolutely safe and secure, wearables are the ultimate solution. Why is this? Wearables know you much better than other smart devices. The main function of the wearable is to collect data from the user. After being processed and passed back, the data can become a unique verification code. In other words, identity verification based on wearable devices could not only identify a person through individual biological features, but also generate a unique and irreplaceable identity code that is derived from a set of specific and abstract data including heart rate, blood pressure, lipids, facial features, skin features, personal preferences, etc. This is exactly why wearable technology is so incredibly fascinating. No other application is as perfectly fitting as this one.

We are living in an age when everyone is worried about privacy, identity theft, and fraud. Data safety has become a major source of concern for individuals and businesses alike. Nobody can be totally free from safety concerns while enjoying the conveniences of the mobile internet. That is why we all have to remember so many passwords and PIN codes. In fact, if you have used Alipay, you would know that every link is interlocked. Even the virtual keyboard for typing in the passcode is encrypted. Users are incredibly concerned about safety.

1. Security checks based on wearable devices

How can identity verification, the killer application for wearable devices, best interact with public life? In my view, wearable devices will become the core to support all future public administration with its unique identity verification function. It will become so crucial that, without this, almost nothing will work in the future. Right now, we can already see the huge obstacles faced by various sectors due to the difficulty of identity verification.

In Shenzhen, one of China's Special Economic Zones, local officials report that security equipment just for the Shenzhen Metro costs RMB 120 million (USD 17 million) even before maintenance costs are considered. Each month the labor cost is over RMB 5 million (USD 714,000), for an annual cost that exceeds RMB 60 million (USD 8.5 million). When the depreciation of equipment is taken into account, the annual cost of security screenings, just for the Shenzhen Metro, is approximately RMB 100 million (USD 14 million). The total revenue of the Chinese Football Association Super League in 2013 was around RMB 220 million (USD 31.4 million), with a net profit of RMB 100 million. The annual cost for security at the Shenzhen Metro was the equivalent of the annual net profit of the Chinese Super League. This is just for the Metro. There are other public transport hubs including airports, bus stations, railway stations, etc., all requiring various levels of security. These places all use traditional security checks, which are labor intensive and time consuming for all. The potential market for wearable-based security checks in China is enormous.

On March 25, 2014, Spring Airlines was the first to successfully board passengers wearing smartwatches using QR code boarding passes. Ticket verification, security checks, and boarding were all completed swiftly.

By placing your wrist close to the ticket-check machine, the process can be completed within seconds. And the same is true for boarding a bus: no need to fumble in your bag anymore, looking for your public transit pass. Just a tap of the wrist is all you need. Using a smartwatch to take public transport is already popular in Beijing.

In public, the first important role for wearables will be in security checks for public transport. When such wearable-based identity verification is developed, it would bring incredible convenience to travel. No more transit cards for buses or subways, no more flashing your not-so-flattering photo ID to strangers. A wrist band will get you through all these gates.

From public transport security checks to other scenarios, this technology will free us from all those membership and bank cards, once for all. In the future, people will no longer need to carry dozens of cards in their wallets. All these cards can be installed into smartwatches. With a single word, the payment interface could pop up. Cash withdrawals or redemption of loyalty card points will no longer require pass codes. Everything can be done with a single tap. Even if the owner loses the wearable device, it would not be a problem as it would provide no use to anyone else. Without the unique body signature of the owner, the watch will not work at all.

Having a wearable device will likely make life much easier for the user and others. A New York studio envisioned how wearables could make public transport more convenient. They conceived a wristband they called Relay, which could access subway data and display the data to its user. For example, when you have to make a decision on whether to take a taxi or the subway, Relay can give you the information on the next train to arrive at the station nearest you.

After analyzing data, a wearable device can inform its user of the best option at any given location, along with the best solutions. It can process information quickly and offer practical solutions. In the process of building smart cities, such technology may become an effective way to increase public transport usage, by making the commute much easier and more pleasant.

This will only be the beginning of wearable technology in public services. We will soon witness wearable devices used in other areas of public administration. When a citizen's data and credit records are planted into the wearable devices, there will no longer be any need for ID cards and the associated risk of losing them. We won't need tedious identity check procedures when passing through security, or the arguments over real-name systems. This will not only save public administration costs, but also increase efficiency and effectively prevent crime.

2. Real-name systems enabled by wearable devices

When you appear online as yourself, would you dare to spread rumors or act irresponsibly? I bet you wouldn't. We have seen various tickets, cards, mobile accounts, and even mobile phones that are now set up under people's real names. How long will it be before the real-name system is applied to other areas? The online environment is particularly in need of regulation. Many have taken advantage of anonymous virtual identities to speak without thinking of the consequences. People abuse others online, rant, and spread rumors, creating much frustration for the "internet cleaners." So far, there has not been any effective way to solve, or even improve, this situation. That is because external pressure only makes the online abusers behave even worse. A real-name system is probably the only solution that could work.

On March 16, 2013, Sina and Tencent both began to apply a real-name system for Weibo, requiring users to provide personal ID information to register. Setting one's front-end user name is voluntary, but back-end registration is based on real name ID verification. Since then, users with no ID verification could only browse without the ability to blog or reblog. In 2015, the State Internet Information Office in China implemented rules for identity information management. Under the principle of "real name in the back end, voluntary use of real name on the front end," Weibo, Baidu Tieba, etc. all adopted real-name identification policies to reinforce online monitoring and administrative enforcement.

The discussion here will not go into detail about the pros or cons of a real-name identification policy, but we can be sure of one thing: through the implementation of a real-name identification policy, the labor and material costs will rise for online social platforms. Take Sina Weibo for example. After adopting the real-name identification policy, three types of costs have gone up. First, the complexity of the system increased, requiring the installation of more servers; second, the website is required to check users' ID information; third, the need for privacy protections increases as do the difficulties, leading to higher operational costs for the website.

There are two types of "citizen ID information checks" available for websites in China, on offer at different price points. One is for personal users, costing

RMB 5 (USD 0.71) each time. The other is for corporate users (e.g. payment websites), charging annual fees and transaction costs cheaper than RMB 5. The annual fees are equivalent to RMB 0.5–1 (USD 0.07–0.14) per ID check on average. According to the statistics published by Sina in 2013, there were already over 500 million Weibo users. At a price of RMB 0.5–1 per check, the real-name identification market for Sina Weibo alone would be worth around RMB 250–500 million (USD 35–71 million).

Wearable technology will accelerate the implementation of real-name systems at full scale, with public security fully covered. As mentioned above, this function can be seen as the killer app of wearables, and the need for identity verification is a significant driver for the emergence of this technology. Presently, when users register their ID number with Weibo, back-office staff still need to manually verify the information to complete the registration. But wearables can help this process happen instantly. When we register with any online social platform in future, the validated identity information could automatically be passed to the operator in less than a second. And the authenticity of the information would be 100% guaranteed, as any identifier would only work with the wearable worn by the specific user. If the wearable were removed from the user, the identification function would be temporarily locked down. Currently no other security measures can match this.

It is difficult to get full insight into the number, due to corporate confidentiality, but we can tell from Sina's example that the real-name identification policy is an open sea with massive potential for exploitation in the future. On one hand it will restrict individuals in terms of speaking in a public space, feeling responsible for what they say, hence making for a more orderly online environment. On the other hand, the government will spare no effort to assist in implementing such an environment.

20.2   Curbing Criminal Activities

According to media reports, police in Dubai have started using Google Glass to identify stolen vehicles. This device has two apps, one of which allows users to record traffic-related offenses, and the other one which allows for the identification of stolen vehicles by examining license plates. Police forces in New York, Los Angeles, and Byron City are also engaged in pilots using Google Glass.

The case of George Floyd in Minneapolis, Minnesota, has attracted a great deal of public attention throughout the entire world. One of the latest of many negative encounters between white police officers and African Americans in the US, there has since been an international uproar after Floyd, an unarmed black man, was killed by a white police officer who suffocated him to death.

Over the past several years it has become a requirement for law enforcement to wear body cameras to capture all encounters with the public. However, despite this there have since been many instances of the officers turning these cameras off before approaching others. There have also been some instances of footage being withheld from the public. Though body cameras were initially meant to put an end to racist or indecent encounters, such instances as those described above have remained a subject of controversy and scrutiny. Wearables are particularly useful in helping police investigate cases. For example, head-mounted polygraphs could be useful in detecting suspects' lies. We could install sensors in a helmet to detect changes in brain waves or nervous system responses when the suspect is presented with pictures of the crime scene or the victim. If the suspect lies, no matter how good their poker face is, the deception can be detected and called out. This would be genuine mind-reading. It is direct information acquisition, rather than interpretation through micro expressions or body gestures. Our thoughts would be more visible than ever with wearable technology.

If a suspect could be quickly identified among a group of ordinary people, this will also enable criminal cases to be solved much quicker. In Brazil, the police force uses a type of smart glasses. They are able to scan 400 faces per second, from up to 50 yards away. The images are compared with photographs in police records on the basis of 46,000 biometric identification points. Once a match is found, the glasses can highlight a suspect with red lines, avoiding spot ID checks all together.

It's clear that different wearable devices, including smart helmets and smart glasses, will play an increasingly important role in assisting law enforcement. The use of wearables will improve investigation efficiency, while also deterring criminality, contributing to a safer and more peaceful society.

20.3   Your City Built by You

The impact of the environment on human health, and the potential role of wearables, is perhaps best illustrated by the case of David Fairley. One day in 1998 Fairley, a statistician by training, was walking down a busy street in San Francisco to pick up his son from preschool, when he suddenly felt weak and dizzy. On the way to the hospital, he suffered from a heart attack.[14]

Fairley partly attributed his heart attack to air quality. After years of investigation, he confirmed previous findings by a researcher in London on the relationship between particles in the air and an increase in a city's death rate, particularly deaths linked to cardiovascular and respiratory system problems.

Even though there were, of course, other factors, I believe walking up that street might well have contributed [to my heart attack]," he said. "Ultrafine particles are so small that they're fairly unstable. They don't stick around. They agglomerate into bigger particles or else diffuse out. If you look at the gradient from roads, concentrations of ultrafine particles are really a lot higher on a busy street. So, it really could make a difference to move one or two streets over.

After many years, David Fairley came to the realization that if one cannot change the environment, he should change his own route, choosing a street with a lower volume of traffic.

In this anecdote we can see the effect of environmental factors on human health. Even though he was aware of this, Fairley was still slow to act. How can wearable technology help in this case?

We are able to install wireless sensors into anti-pollution facial masks which could collect real-time data from the streets, e.g. particle concentrations in the air, carbon dioxide density, PM2.5 levels, etc. This can be compared with historical data from governments and academic institutions. It would then be able to send correct information to the user in time to prevent any issues from happening.

If David Fairley had one of these devices, he would not have waited years before changing his walking route. What could have happened was his wearable device would have immediately sent him alerts warning that particle concentration levels on that street would pose potential health risks or even trigger heart attacks. It also could have provided recommendations for alternative routes. This would have given Fairley better information with which to make a proper decision on his route in time to avoid being exposed to adverse environmental factors.

1. Working together with governments and academic institutions

More and more wearable technology companies are providing data for academics and governments. When the air quality index reaches a certain level, the wearable device will warn its user, informing them of the exposure and potential health risks. With such information, air pollution control agencies could issue more rational standards and create more effective policies.

The main technical challenges lie in the analysis of Big Data and the development of standards, which are also massive business opportunities for many wearable technology companies or data analysis firms. The core of wearable technology is the collection and deep processing of data. Without analysis and feedback, data collection alone does not bring much value or revenue.

By providing analytic reports to government agencies, we could witness incredible benefits regarding the design of urban environments in the future. Accuracy and efficiency will be improved significantly in planning, policy making, and public engagement.

There are companies already exploring opportunities in this area. Jonathan Lansey is a data engineer at Quanttus, a wearable technology company.[15] This company is developing a new watch that can measure and analyze users' vital signs—including heart rate, blood pressure, and body temperature. This data can be used to assess how they respond under different conditions, including different air pollution levels or various weather conditions. The company is also seeking to partner with academic institutions, hoping to provide data for research studies. Lansey has said that Quanttus plans to provide user data to academic institutions. By using this data, hopefully, scientists can gain more insight into the impact of environmental factors on human health, and provide supporting evidence for government policymaking.

"I think that's going to be a large branch of our contribution; we're building a business here, but there's an altruistic component, too," said Steve Jungmann, vice president of product management for Quanttus. "We're across the street from MIT, so we get to talk to people there with some fantastic ideas around biometric monitoring as attached to x, whether it be air quality, emotions, the ability to perform under pressure, any number of things," he said. The company is also making devices for various research purposes.

2. Smart cities

In June 2015 Larry Page, CEO of Google, published a post on Google+ explaining that the company would establish a new urban innovation company called Sidewalk Labs, to improve the lives of billions of people on the planet. Sidewalk Labs focuses on developing new products, platforms and collaborating partnerships, to solve problems in living, transportation, and energy usage. Google has a clear and simple company positioning: "people." The aim is to make people more comfortable in a city, freeing them from concerns about traffic congestion, parking availability, long queues for food, and poor air quality. How to achieve these objectives? One solution is to make the city around us more intelligent.

IBM defines a smart city as a city that fully utilizes all available internet information for better understanding and control of city operations, as well as the optimal use of limited resources. Technology will be the building blocks of future cities. Without technology, cities will go nowhere. As one of the forerunners for the smart cities of tomorrow, IBM has provided solutions for many countries and cities in terms of urban transportation management options.

By studying over 50 cities in developed and developing countries, IBM has found that all of the cities in the world have their own transportation issues. But some cities have achieved significant results with the help of IBM Smart Transport solutions. Take Stockholm, for example. As the capital of Sweden, the city accommodates over 500,000 vehicles every day. In 2005, the average commuting time was 18% longer than the previous year. The Royal Swedish Academy of Sciences then began a collaboration with IBM to develop and implement a smart transportation system that was suitable for the area. In 2006 this smart transportation system was adopted, and in the following three years traffic congestion was reduced by 25%. Time spent waiting in traffic also was reduced by 50%. Taxi revenue increased by 10% and urban pollution dropped by 15%. Stockholm also saw a rise of 40,000 new public transit passengers.

How smart a city is cannot be reflected by traffic alone. Traffic is something on the surface of a city. Technology is what's underneath and wearable technology can prove most useful in areas like healthcare and education. Wearables not only bring a hands-free experience but, more importantly, redefine our lifestyle. In the future, omnipresent mobile internet can connect wearable devices, to make distance learning, telemedicine, and taxi businesses all possible. With the rapid development of the Internet of Things and smart cities, information flows will be more open and interactive in the future. Entire cities could be felt and understood on a different level. New government proposals, open for public comment, will be distributed to wearables through dedicated systems that welcome your comments and feedback. You could be participating in these matters before you know it.

You will be contributing to the planning and construction of your city, sharing issues and ideas with fellow citizens. At the same time, you can also be kept informed about the progress of each plan. At the moment, such privileges are granted through applications to certain government units with a lot of necessary paperwork and justification.

In the future, unified platforms may provide a space for sharing information and interaction between citizens, policymakers, and implementing authorities.

In cities of the future, all urban management will be established upon a massive and complete touchscreen-based interactive map. Urban managers only need to drag and click to complete the configurations. If you have watched the movie *The Hunger Games,* you can imagine what it may look like. Having solutions at your fingertips, no matter what your needs are, is part of what wearable technology can provide.

NOTES

1. Figures taken from Mobile World Congress (2015) *The Mobile Economy 2015*.
2. "Is Nike FuelBand out of gas Already"
   https://www.cnet.com/news/is-nike-fuelband-out-of-gas-already/
3. Christina Farr. "Wearables with a Purpose: Emotiv Rallies the Community to
   Build a Mind-Controlled Wheelchair." https://venturebeat.com/2014/01/20/
   wearables-with-a-purpose-emotiv-rallies-the-community-to-build-a-mind-
   controlled-wheelchair/
4. China Internet Network Information Center (CINIC)
5. "Abstract Games with Raph Koster" https://www.raphkoster.com/games/
   interviews-and-panels/abstract-games/
6. "Mark Zuckerberg: Here's Why I Just Spent $2 Billion On A Virtual-Reality
   Company" http://www.businessinsider.com/zuckerberg-why-facebook-bought-
   oculus-2014-3?IR=T
7. "Bing Gordon: Every Startup CEO Should Understand Gamification" https://
   techcrunch.com/2011/06/30/bing-gordon-every-startup-ceo-should-understand-
   gamification/
8. "Quick, Wearables, Hide! The Ads Are Coming…" https://venturebeat.
   com/2014/07/16/quick-wearables-hide-the-ads-are-coming/
9. "Advisers Target Wearable Gadgets as Next Ad Frontier" http://www.inmobi.com/
   company/news/advertisers-target-wearable-gadgets-as-next-ad-frontier/

10. "Companies Hunt for the Next Ad Frontier" https://www.iol.co.za/business-report/companies/hunt-for-the-next-ad-frontier-1715734

11. "Could Wearable Tech Read Minds to Sell Ads?" https://www.cnet.com/news/could-wearable-tech-read-minds-to-sell-ads/

12. "Device That Can Literally Read Your Mind Invented by Scientists" https://www.independent.co.uk/news/science/read-your-mind-brain-waves-thoughts-locked-in-syndrome-toyohashi-japan-a7687471.html

13. "Neuromarketing" https://en.wikipedia.org/wiki/Neuromarketing

14. "Wearable Tech Helps You Live in the Moment" https://www.scientificamerican.com/article/wearable-tech-helps-you-live-in-the-moment/ Citation needed...

15. "The Future of Wearables Makes Cool Gadgets Meaningful" https://www.theatlantic.com/technology/archive/2014/05/the-future-of-wearables-makes-cool-gadgets-meaningful/371849/

BIBLIOGRAPHY

[1] Wearable Market Half-Year Analysis Report 2016 GfK China
[2] Wearable Market Shipments Q1 2016, Market research firm IDC
[3] Data source regarding Fitbit, the Apple Watch, and other products, Slice Intelligence

INDEX

INDEX

ABOUT THE AUTHOR

Dr. Kevin Chen, or Chen Gen, is a renowned science and technology writer and scholar, postdoctoral scholar at Boston University, and an invited course professor at Peking University. He has served as special commentator and columnist for *The People's Daily*, CCTV, China Business Network, SINA, NetEase, and many other media outlets. He has published several monographs involving numerous domains, including finance, science and technology, real estate, medical treatment, and industrial design. He has currently taken up residence in Hong Kong.

CPSIA information can be obtained
at www.ICGtesting.com
Printed in the USA
LVHW092154070521
686849LV00001B/1